BREADFRUIT

Arts and Traditions of the Table: Perspectives on Culinary History

ARTS AND TRADITIONS OF THE TABLE:
PERSPECTIVES ON CULINARY HISTORY
Albert Sonnenfeld, Series Editor

What We Eat: A Global History of Food, edited by Pierre Singaravélou and Sylvain Venayre, translated by Stephen W. Sawyer
Spoiled: The Myth of Milk as Superfood, Anne Mendelson
The Fulton Fish Market: A History, Jonathan H. Rees
The Botany of Beer: An Illustrated Guide to More Than 500 Plants Used in Brewing, Giuseppe Caruso
Anxious Eaters: Why We Fall for Fad Diets, Janet Chrzan and Kima Cargill
Gastronativism: Food, Identity, Politics, Fabio Parasecoli
Epistenology: Wine as Experience, Nicola Perullo
The Terroir of Whiskey: A Distiller's Journey Into the Flavor of Place, Rob Arnold
Meals Matter: A Radical Economics Through Gastronomy, Michael Symons
The Chile Pepper in China: A Cultural Biography, Brian R. Dott
Cook, Taste, Learn: How the Evolution of Science Transformed the Art of Cooking, Guy Crosby
Garden Variety: The American Tomato from Corporate to Heirloom, John Hoenig
Mouthfeel: How Texture Makes Taste, Ole G. Mouritsen and Klavs Styrbæk, translated by Mariela Johansen
Chow Chop Suey: Food and the Chinese American Journey, Anne Mendelson
Kosher USA: How Coke Became Kosher and Other Tales of Modern Food, Roger Horowitz
Taste as Experience: The Philosophy and Aesthetics of Food, Nicola Perullo

For a complete list of books in the series, please see the Columbia University Press website.

Breadfruit

THREE GLOBAL JOURNEYS
OF A BOUNTIFUL TREE

Russell Fielding

Columbia University Press
New York

Columbia University Press
Publishers Since 1893
New York   Chichester, West Sussex

Copyright © 2025 Columbia University Press
All rights reserved

Library of Congress Cataloging-in-Publication Data
Names: Fielding, Russell, 1977– author
Title: Breadfruit : three global journeys of a bountiful tree / Russell Fielding.
Description: New York : Columbia University Press, [2025] | Series: Arts and traditions of the table : perspectives on culinary history | Includes bibliographical references and index.
Identifiers: LCCN 2025010375 (print) | LCCN 2025010376 (ebook) | ISBN 9780231219075 hardback | ISBN 9780231219082 trade paperback | ISBN 9780231562539 ebook
Subjects: LCSH: Breadfruit—History
Classification: LCC SB379.B8 F54 2025 (print) | LCC SB379.B8 (ebook) | DDC 634/.39—dc23/eng/20250804

Cover design: Milenda Nan Ok Lee.
Cover art: Margaret Girvin Gillin, *Breadfruit*, ca. 1880. Bishop Museum Archives.

GPSR Authorized Representative: Easy Access System Europe, Mustamäe tee 50, 10621 Tallinn, Estonia, gpsr.requests@easproject.com

For Conrad and Margaux

Contents

Preface ix
A Note About Language xv

PART I The Pacific

I The Giving Tree 4

II Charismatic Megaflora 23

III The Treeness of Life 39

IV Awful and Lovely 55

V The Wondrous Food of the Land 71

PART II The Caribbean

VI A Dish de Résistance 92

VII An Idea Whose Time Had Come 106

VIII No Poem About Breadfruit 116

IX Fruits of the Creole Kind 131

X Roast or Fry 143

PART III   The World

XI   The Second-Best Time to Plant a Tree   164

XII   Super Food   181

XIII   Pushing Latitude   196

XIV   Two Trees   219

Acknowledgments   233
Notes   237
Index   277

Preface

The first time I tasted breadfruit was in a meal purchased from a pop-up food stand at the corner of two unpaved roads on the island of St. Vincent. I had been in the Caribbean for several months, conducting research on fisheries and marine conservation. I always looked forward to these weekend street-corner barbecues for the change they offered from the wonderfully fresh but repetitive fish I would often bring home from the boats on which I was working. Pork ribs were the special on this day, and as the chef was tonging the meat from the grill into a clamshell Styrofoam container he asked me about a side dish: "macaroni pie or provisions?"

Provisions, or "ground provisions" as they're often called, are a hearty mix of mostly root vegetables, seasoned and steamed or boiled together. They're also an echo of the Caribbean's past. During the plantation era, cash crops, mostly sugar, would be planted, tended, and harvested by forced labor five-and-a-half days a week. The enslaved, as the early eighteenth-century traveler Hans Sloan explained, "have Saturdays in the Afternoon, and Sundays . . . to feed themselves from . . . Ground allow'd them by their Masters." In their kitchen gardens and provisioning grounds, enslaved Caribbean plantation laborers grew food for their own consumption. These were their provisions, their sustenance as they worked against their will to make other people rich.[1]

Figure 0.1 Provisions, including breadfruits at center and lower right, for sale at a market stall in St. Vincent and the Grenadines. (Photograph by Russell Fielding.)

"Um, provisions," I told the chef, who quickly scooped a heap of vari-colored starchy slices onto my dish. He must have seen the confusion on my face as I confronted these unfamiliar vegetables. "That's dasheen, that's pumpkin, that's yam," he explained, gesturing with the grill tongs, "plantain, breadfruit." This last word stuck in my mind as I walked back to the little apartment, attached to a family's home, that I'd rented during my research fieldwork: *breadfruit*.

I had heard of this, this—fruit?—before, and had enjoyed the welcome shade of its trees, but I realized that I must have imagined something entirely different as a food. I sat down on the porch and considered the grayish-white, crescent-shaped slice lying alongside the other vegetables in the container. It appeared dense and starchy. I took a bite. Having been boiled with the rest of the provisions, the breadfruit had absorbed some of the taste, and a little color, from its pot-mates: particularly the flavorful plantain and the carotene-rich pumpkin. Its texture was at once chalky and spongy; it was neither juicy nor completely dry. It seemed like something was missing—not from the breadfruit, but from my first impression of it. I could tell there was more than what I was experiencing. For the rest of my fieldwork season I sought out breadfruit whenever I could find it on a menu or at a market.

Teaching geography and environmental studies at the university level, I kept a regional research focus on the Caribbean, particularly St. Vincent and its nearby islands, taking on projects ranging from fishing to whaling to renewable energy development and seawater desalination. Breadfruit was present, in the background, during all of this. It wasn't until much later that I decided to make it the central subject of my research.

When a sabbatical year became available in mid-2019, my family and I moved to Barbados, where I had a visiting appointment as a lecturer at the University of the West Indies–Cave Hill. We saw breadfruit every day: stacked neatly at the Bridgetown fruit and vegetable markets, on display at local grocery stores, and as a side dish at most local restaurants. The stately trees with their huge lobed leaves were everywhere, and the large green fruits would occasionally fall from a branch and roll down the street in our hilly neighborhood. We ate it often, and I began to collect information for what was becoming a new research project.

Then one early spring day in 2020 we heard news of social gatherings being restricted, restaurants shifting to takeout only, schools and universities moving to online instruction, and international borders being closed.

Cautious, I booked four tickets on a flight to Miami scheduled for late April. Then a representative from JetBlue Airlines called me one Friday in early March and said our flight had been canceled. *All* flights had been canceled, but there was one departing the next morning with four seats available. We could either change our tickets and fly out the next day or stay in Barbados indefinitely. We packed our things, said quick good-byes to our friends, and took the flight.

Returning to the United States prematurely with no travel possible in my foreseeable future, I dived into the literature, reading everything I could find on the subject of breadfruit. It became clear to me that breadfruit's history could be thought of as comprising two major world journeys: first, its distribution from a point of origin in Southeast Asia out across the vast Pacific, and second, transplanted halfway around the world from the Pacific islands to the Caribbean. I also came to understand that a third major journey was currently taking place, one that would carry breadfruit throughout the world.

Beginning more than three thousand years ago during the great Pacific voyages of discovery, breadfruit took its first long journey: from Southeast Asia across the Pacific Ocean, finding soil and sunshine on nearly every tropical island in the region. Pacific Islanders found myriad uses for breadfruit including but extending far beyond food: ancient Hawaiian surfers carved boards from the buoyant wood of its trunks and sanded them with its rough-surfaced leaves, boatbuilders used its sap as caulk, and artisans pounded its bark into fine cloth. One eighteenth-century naturalist would call breadfruit "the most useful of all the Fruits of the East Indies." Breadfruit still grows widely throughout the Pacific but, on some islands, it has become neglected in favor of imported foods like bread, potatoes, and rice. On some of these islands, though, breadfruit is experiencing a renaissance.[2]

Later, as European colonial governments were busy cruelly engineering the large-scale forced movement of Africans to the Caribbean islands to work their sugar plantations, they sought a food source that required little input of labor and took up minimal acreage to fuel this enslaved workforce. Explorers' glowing reports of breadfruit's easy productivity, along with persistent pleas from colonial planters, convinced the British Crown to finance two round-the-world voyages to bring breadfruit to the Caribbean from the Pacific. The first ended in failure and mutiny, but in 1793 the second voyage successfully brought hundreds of potted young breadfruit trees from Tahiti to St. Vincent, completing breadfruit's second long journey. Trees were distributed throughout the Caribbean islands, where they soon began to produce fruit. The enslaved laborers detested breadfruit, which they considered unpalatable, bland, and far inferior to the other provisions they had grown used to. After emancipation the tree stood as a reminder of slavery and remains a symbol of oppression to some in the Caribbean today, even as others have embarked on a mission to redeem breadfruit, like the other things of value they have reclaimed from the wreckage of empire.

In recent years, organizations involved in sustainable development have begun to consider breadfruit a worthy ally in the fight against both world hunger and environmental degradation. As a tree crop with a lifespan of many decades, breadfruit captures carbon from the atmosphere and stabilizes soil, countering both climate change and erosion—all while providing a reliable source of food for humans and habitat for other species. To this end, organizations like the Chicago-based Trees That Feed Foundation have begun breadfruit-planting programs throughout tropical Africa, Asia, and Central America. A 2016 article in the culinary magazine *Saveur* asked rhetorically, "Can breadfruit save the world?"[3]

Simultaneous to its emergence on the food security and sustainability scenes, breadfruit appears to be gaining recognition as a "superfood" within communities focused on wellness and plant-based nutrition. Praise for breadfruit can be found in media ranging from *Vegan Review* to the *Wall Street Journal*. Most mention its nutritiousness as well as the fact that it can be milled into a white, gluten-free flour. *Bon Appétit* and *Food & Wine* magazines, the James Beard Foundation, and the Food Network have all recently featured breadfruit recipes. Author Andrea Gibbons has called breadfruit "*the* new vegetarian alternative." *Forbes* magazine and the *New York Times* have both predicted breadfruit's emergence as a "future food trend." As a way to feed the hungry while restoring the environment, and as a trendy superfood, breadfruit's third long journey is taking it all over the world.[4]

When the COVID-19 lockdown began to ease, field-based researchers like me started to travel again, domestically at first. I made several research trips to Hawai'i, where breadfruit has grown for hundreds of years, and to Florida, where a warming climate is just beginning to make it possible at scale. Then, in early 2022, I received a grant from the Alfred P. Sloan Foundation, which allowed me to increase my research fieldwork dramatically just as international travel was reopening.

Thanks to the Sloan Foundation's funding, I have made several long, island-hopping trips across the Pacific, visiting breadfruit growers and those who use it for food and other resources in Tahiti, Moʻorea, Bora Bora, Maupiti, Tetiʻaroa, and the Marquesas in French Polynesia; Pohnpei in Micronesia; Fiji; Sāmoa; and the U.S. territories of American Samoa, Guam, and the Northern Mariana Islands. I returned several times to the Caribbean, spending time with breadfruit growers, processors, chefs, and consumers in Barbados, Grenada, Puerto Rico, St. Barthélemy, St. Lucia, St. Vincent and the Grenadines, the U.S. Virgin Islands, and, though they aren't geographically part of the Caribbean, I followed breadfruit to the Bahamas and Bermuda too.

To see the results of sustainable development work involving breadfruit, I traveled to Ghana to ride along with an agricultural extension officer on his rounds, visiting farmers growing donated breadfruit trees and vendors selling their breadfruit-based products. As in Florida, Australian fruit growers are finding that climate change makes it easier to grow breadfruit there; I learned about this firsthand from farmers I visited throughout Queensland. I traveled to botanical gardens and archives, and spoke with experts in Chicago, London, New York, Miami, and Bali. Finally, I followed breadfruit to the place of its evolutionary origin on the island of Borneo in Malaysia.

This book follows a course not unlike that of the three global journeys taken by breadfruit itself, across the Pacific, to the Caribbean, and then around the world, exploring and learning from science, literature, and the everyday experiences of people I met. My own ancestry traces neither to the Caribbean nor to the Pacific islands. As a regular visitor to these two great regions of breadfruit's history, it is important to me that this book presents and amplifies the actual voices of people with direct, lived experience of breadfruit, not just a filtered summary compiled by a distant observer. Still, what I offer here is my own interpretation of the legends, science, traditions, and wisdom that were shared with me. I hold no original knowledge of breadfruit, and it is only because of the generosity of those who do that I have learned what I have come to know. I feel immense gratitude to everyone I met along these journeys—everyone who shared their knowledge and often their food with me during my travels.[5]

A Note About Language

The tree and fruit that we call in English "breadfruit" goes by a variety of names in the places around the world where it grows. Most European languages use a term that is simply a calque—a direct translation—of the name bestowed by the seventeenth-century English pirate and explorer William Dampier: *bread-fruit*. Consider the Dutch *broodvrucht*, the German *brotfrucht*, the French *fruit à pain*, the Portuguese *fruta-pão*, and the Spanish *frutapan*. Around the year 1900 the English term seems to have lost its hyphen, and it became simply *breadfruit*.

Since the plant itself is native to Southeast Asia and the Pacific islands, it is worth giving special consideration to the names used in those regions. In Indonesia and Malaysia, where its ancestors evolved, breadfruit is called *sukun*. In New Guinea, where breadfruit was first domesticated, it is called *kapiak*. Throughout most of the Pacific islands, the term used is a cognate of either *kuru* or *mei*. For example, in the Federated States of Micronesia, Kiribati, the Marshall Islands, the Marquesas, Tonga, and Tuvalu, the terms *mei*, *mai*, or *mahi* are used. CHamoru speakers in Guam and the Mariana Islands use the slightly modified *lemai* or *lemae*. In the Cook Islands, the word is *kuru*; in Hawai'i, Sāmoa, and part of Fiji, it's *ulu*. (Hawaiians add an *'okina*, which resembles an apostrophe, before the first *u*. This symbol signifies a glottal stop, the sound represented by the hyphen in the word *uh-oh*.) Throughout the rest of Fiji, *ulu* is *uto*. In the Society Islands, the archipelago that includes Tahiti, it is *'uru*. Bridging the gap between the *mei* and *kuru* families is the Palauan term, *meduu*. Linguists theorize that most of these terms descend from the proto-Oceanic language's word for the fruit: **kulur*.[1]

For the sake of simplicity, in this book I shall mainly refer to breadfruit as *breadfruit*, especially when discussing its role in the Caribbean—where its common name is usually that of the colonial language of any given island—or in places where breadfruit has been introduced more recently. When discussing scientific research into breadfruit, I will occasionally use the plant's scientific name, *Artocarpus altilis* and will discuss the meaning of this name later. Finally, when dealing specifically with breadfruit's history and current use in the Pacific, I will often use the Hawaiian term *'ulu*, even in discussions that are not explicitly focused on Hawai'i. In doing so, I do not mean to elevate the Hawaiian language above the many other Pacific languages to which it is related. Instead, I mean only to recognize the enormous debt I owe to Hawai'i and to the chefs, growers, producers, promoters, scholars, and teachers of *'ulu* I met there. Of all the places I traveled for this research, it was in Hawai'i that I had the most intensive and meaningful *'ulu* experiences and lessons. My use of the word *'ulu* anytime I refer to breadfruit in the Pacific context is a way of saying *mahalo nui loa* to my Hawaiian *'ulu* guides and teachers.

Several years ago, in an online discussion forum on the travel website TripAdvisor, a user from Arizona asked his fellow travelers for advice on finding fried breadfruit during an upcoming trip to Maui. A Maui-based user writing under the name *kaiwahine* ("ocean woman") responded, helpfully, to say that fried *'ulu* was rare but that one could readily find it steamed or baked. Ocean Woman even suggested a restaurant. A follow-up response from the original poster asked whether the word *'ulu* was "Hawaiian for breadfruit." The reply from kaiwahine set the record straight: "No," she simply and correctly stated, "*breadfruit* is English for *'ulu*."

PART I

The Pacific

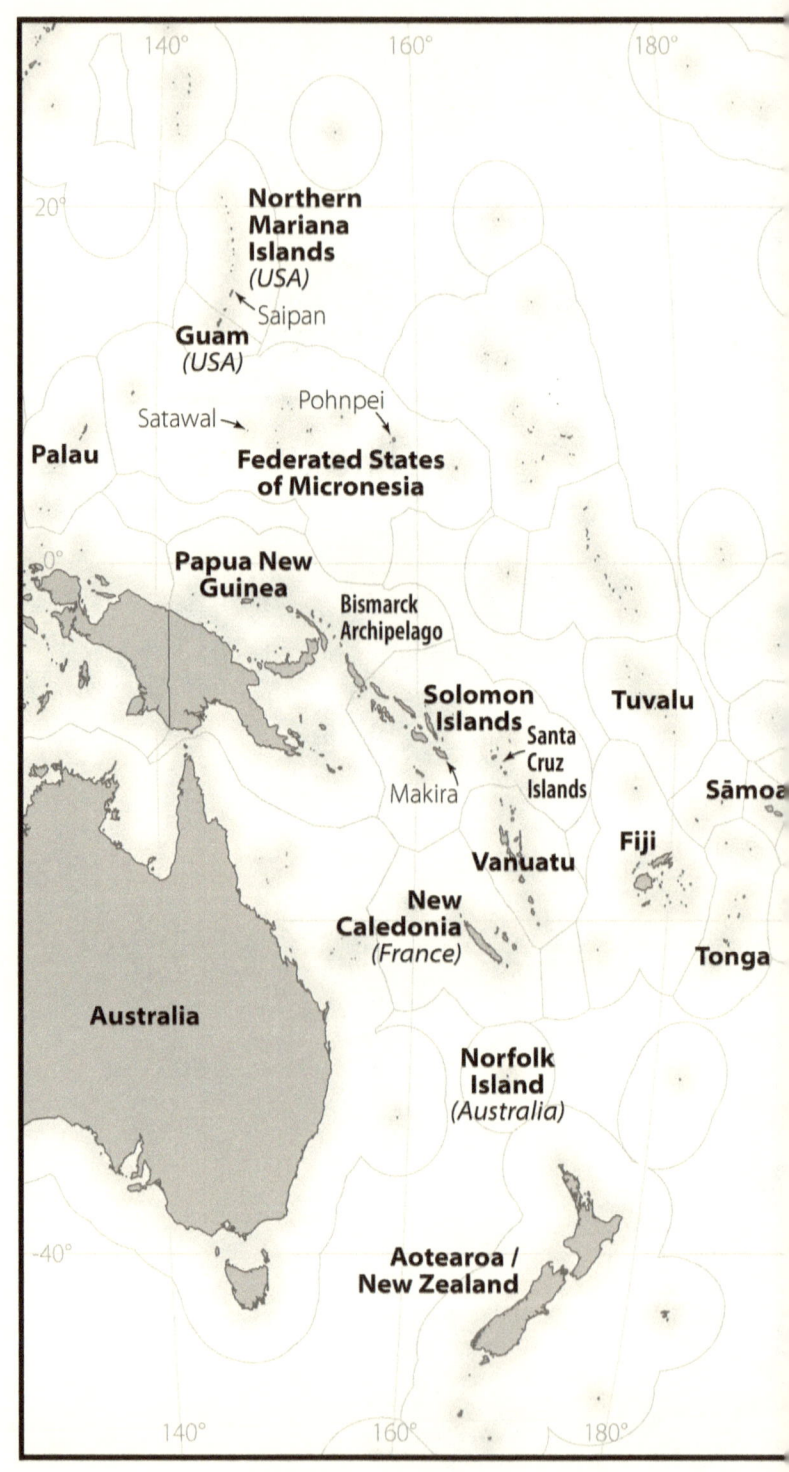

Map 1 The Pacific. (Cartography by Alison DeGraff Ollivierre, Tombolo Maps Design.)

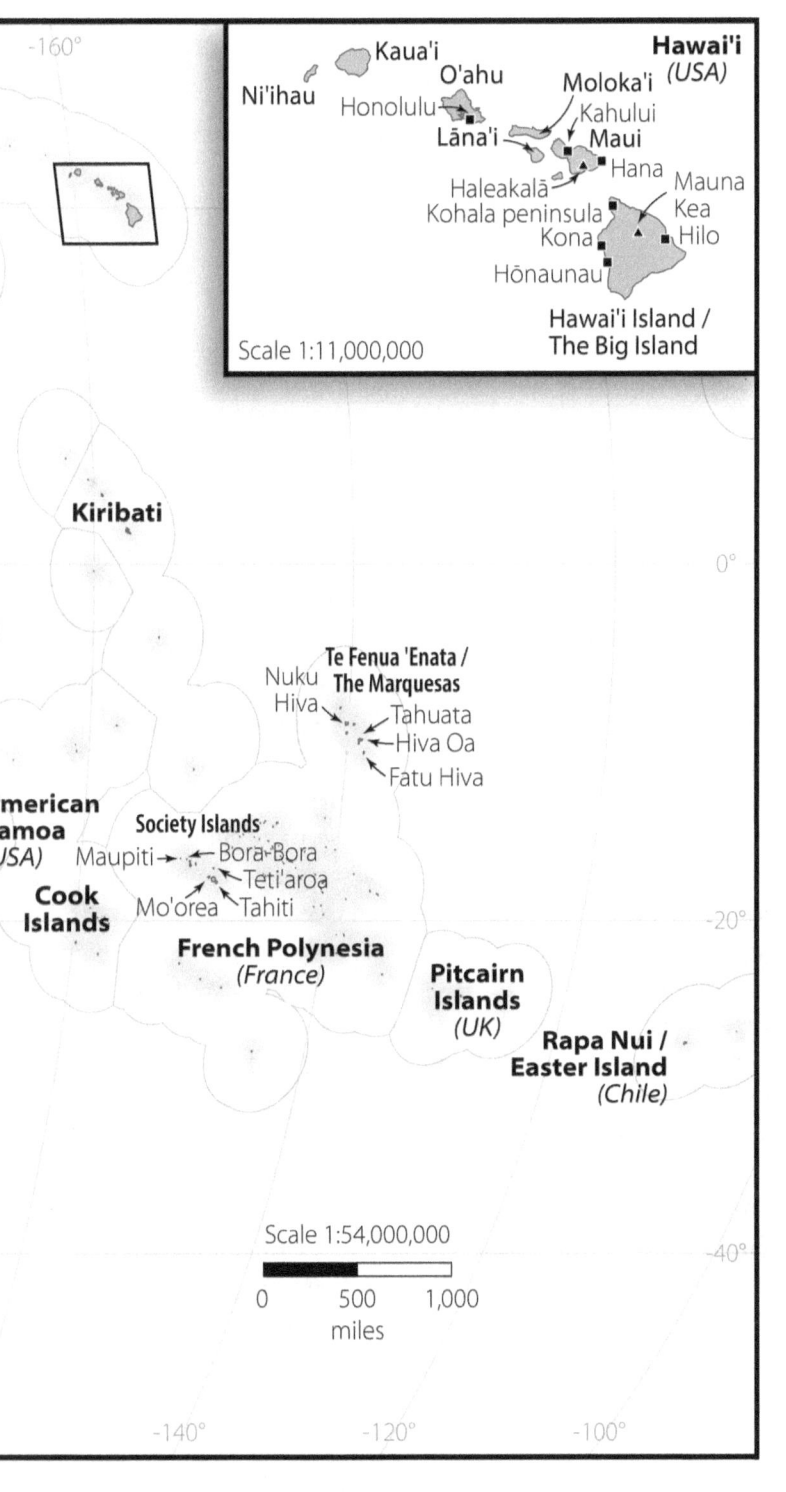

CHAPTER I

The Giving Tree

'*Ulu* is a gift. While details of the story vary as it is told throughout the vast Pacific region, this much is widely agreed upon: *'ulu* traces its origin to an act of self-sacrifice. In Hawai'i, the story is told of the god Kū, who, living in disguise as a mortal man, gave himself up for his mortal wife, Hina, and their children. Hawai'i was in the depths of a famine and Kū, unable to bear watching his *'ohana* succumb to starvation, sunk headlong into the soil and regrew as a tree, watered by Hina's tears. His last spoken words were an admonition to pick, roast, and eat the new and unfamiliar fruit that would grow from his branches.[1]

Elsewhere throughout the Pacific, stories similar in theme, if differing in detail, are told. An alternate version from Hawai'i says it was a mortal, not a god, a man named 'Ulu, who died of starvation and was buried by a stream, only to feed the roots of the first *'ulu* tree that would sprout from his grave, and, in doing so, saved his family from a similar fate. A legend in Vanuatu, tells of a woman who cut off her own breast, which became a breadfruit, to feed her starving young sister. On the island of Ra'iātea, also during a famine, a man is said to have told his wife, "I will die and become food for our son." He instructed her to bury his body and predicted that a tree would arise therefrom, but not just any tree—it would be his own body, reincarnated: "my hands become leaves, the trunk and two branches my body and legs, and the round fruit my head." The man soon died, his wife buried the body, and from the grave a breadfruit tree sprung up, bearing food that saved the lives of the woman and the young boy.[2]

The title character of Shel Silverstein's book *The Giving Tree* provided a boy with apples for eating and selling, branches for climbing and house-building, a trunk for boat-building, and finally a stump for sitting. After giving each gift and seeing how it had helped the boy, the tree was happy. Imagine, then, the joy felt by an *'ulu* tree as a result of the gifts it can give.[3]

Foremost is the nutritious food, which a breadfruit tree produces abundantly. Tusi Avegalio, a traditional Samoan chief and retired University of Hawai'i professor, has called breadfruit "the most prolific fruiting tree in the world." While the exaggerated figure of each tree producing up to seven hundred fruits per year is often cited, Diane Ragone, director emerita of the Breadfruit Institute, has analyzed a number of productivity studies conducted throughout the world, finding the average yield to be closer to two hundred. But farmers in Tanzania, where breadfruit was introduced from across the Indian Ocean in the 1950s, have reported annual yields of more than nine hundred cantaloupe-sized fruits per tree. This may sound fantastical but keep in mind that, after a very long journey with stops lasting at least a generation—from the Pacific islands to the Seychelles to Zanzibar to mainland Tanzania—highly productive individual trees would have been purposefully selected for export, and the clonal offspring of these champions may have found ideal growing conditions in the rich African soil of the Great Rift Valley.[4]

Joseph Banks, a wealthy eighteenth-century gentleman-scientist, as self-funded English scholars were then known, who sailed the Pacific with Captain James Cook aboard HMS *Endeavour*, was impressed with breadfruit's productivity. He quoted Virgil in describing the fortunate lives of the Tahitian people whom he had observed living among an abundance of breadfruit: "*O fortunatos nimium sua si bona norint*" (They would count themselves lucky, if they only knew how good they had it). To Banks, it seems as though bread was simply growing on trees.[5]

The breadfruit trees of Tahiti, of course, being a cultivated crop, did not sprout all by themselves. Significant labor was expended in preparing land, propagating saplings, and tending trees. But the mature groves that Banks, Cook, and the rest of the *Endeavour*'s crew encountered appeared, to them, to provide a near-constant, labor-free source of food. To Banks this observation seemed to have spiritual, as well as agricultural, significance.

He and other Europeans who would come to the Pacific after him conceptualized the Tahitians and their fellow Pacific Islanders as a population still enjoying an Edenic geography, unencumbered by the divine affliction with which Adam and Eve were sent east of their garden with God's imprecation: "cursed is the ground for thy sake; in sorrow shalt thou eat of it all the days of thy life . . . in the sweat of thy face shalt thou eat bread."[6]

In contrast to this curse, Banks wrote of the Tahitian people, "scarcely can it be said that they earn their bread with the sweat of their brow when their chiefest substance Breadfruit is procur'd with no more trouble than that of climbing a tree and pulling it down." He went on to explain that, "if a man in the course of his life planted ten such trees (which, if well done, might take the labour of an hour or thereabouts), he would as completely fulfil his duty to his own as well as future generations, as we, natives of less temperate climates, can do by toiling in the cold of winter to sow, and in the heat of summer to reap, the annual produce of our soil." This productivity amazed Banks, but it also contributed to the common and incorrect view of Pacific Islanders that would persist among Europeans for generations: that they were lazy and did not work because they didn't need to, mostly thanks to the productivity of their *'ulu* trees.[7]

Most Pacific societies have not simply chosen leisure in response to bountiful breadfruit harvests. Like the early agriculturalists who ushered in the Neolithic Revolution in Mesopotamia, many prehistoric Pacific island nations seized the opportunities presented by such an abundant and reliable food supply. With their nutritional needs fulfilled, these societies would have felt release from the day-to-day pressures of subsistence. Especially after techniques of preservation by fermentation were developed, breadfruit would have served to extend a society's comfort zone beyond the short term and to allow greater diversification of labor. Rather than everyone, or nearly everyone, being engaged in food production, some members of a breadfruit-rich society could specialize in the arts, or government, or religion, or war. Socioeconomic classes could emerge. Culture could flourish. A revolution could begin.

Anthropologist Glenn Petersen presented the history of what he has called "Micronesia's breadfruit revolution," a transformation of island landscapes from native forest to agricultural fields, and a subsequent transformation of people's way of life from simple to highly complex. Petersen has highlighted the Micronesian island of Pohnpei as the one place where the effects of this breadfruit revolution are most apparent, particularly focusing upon the archaeological site of Nan Madol. This massive stone city was built between four hundred and a thousand years ago during a time of political and economic organization made possible, in part, by bountiful harvests of breadfruit.[8]

Nan Madol may be the house that breadfruit built, but today it is like a grand house whose family has gotten too small to inhabit its entirety. Like an old couple whose children have grown up and moved out, the Pohnpei Historic Preservation Office maintains for visitors only a small portion of what once was a massive megalithic city. The large stones that comprise the structures have led some observers to call Nan Madol "the Stonehenge of the Pacific" and the canals interlacing the city's many artificial islands have lent it the name "Venice of the Pacific." But walking and wading through this complex, my mind drifted neither to the Salisbury Plain nor to the Laguna Veneta. Instead, I found myself agreeing with the physician and naturalist Oliver Sacks, who wrote that "there is nothing on the planet quite like Nan Madol."[9]

I had arrived at Pohnpei on a late flight from Guam, touching down at the stroke of midnight on New Year's Day 2023. The next morning my guide, Eugene Darsy, picked me up at my hotel in Kolonia, the island's capital. As we left the city and began to cross the island toward Nan Madol, Eugene said, "this side, you will see many things; other side of Pohnpei," meaning the side where the capital was located, "nothing interesting."

We left Eugene's car at the end of a dirt road and followed a faded, hand-stenciled sign pointing toward the Nan Madol Trail. As we walked, some especially tall breadfruit trees came into view. Eugene had already mentioned the years he had spent in the United States, studying at the University of Oregon, located—appropriately—in Eugene. Walking barefoot under the tall breadfruit trees, he told me, "I wore shoes a lot when I lived in the States. Now, when I try to climb trees, my feet get tickly. If you have ticklish feet and you climb a breadfruit tree, the tree will die. If you're scared of heights and you climb, the tree will die." I noted this warning and resolved to climb breadfruit trees only when I was feeling neither ticklish nor acrophobic.

Continuing down the trail, we passed a house where young men sat pounding *sakau*, a local Micronesian version of what is called *kava* in other parts of the Pacific. This root produces a mildly sedative drink that looks, and tastes, like thin, watery mud. *Sakau* contains no alcohol but fills a similar social role in Pohnpeian culture as alcoholic beverages do elsewhere: it

THE GIVING TREE [7]

relaxes people and tends to lubricate social interactions. Pharmacological researcher Louis Lewin wrote in the early twentieth century of *sakau*'s effects: "a state of happy carelessness, content and well-being appears without any physical or mental excitation." Lewin emphasized that "those who drink it are never . . . angry, aggressive and noisy, as in the case of alcohol," but that "reason and consciousness remain unaffected."[10]

The effort of the young men wielding stones was in preparation for a party planned later that afternoon. After pulverizing the *sakau* roots with baseball-sized rocks, they would add water and twist the resulting sludge inside ribbons of stripped hibiscus bark before straining it out into drinking vessels. "Come back later for *sakau*," the stone-pounders implored us as we passed by on our way toward Nan Madol. We assured them we would and walked on to the *ting-ting-ting* rhythm of their stones striking the flat tabletop of basalt upon which they had spread the precious roots.

As we approached the section of Nan Madol that's kept up for visitors, I began to notice lengths of columnar basalt strewn alongside the path or sunk to the bottom of the crystal-clear canals lining the elevated trail. Most of these cast-offs were longer than I am tall and wider than my torso, but when I found a short piece, maybe half a meter in length, I tried to lift it. No chance. It was far too heavy even to roll.

With this basaltic heaviness in mind, Eugene and I crossed a flimsy log bridge and came out of the forest. The sun shone on the clear water of a canal separating us on one side from small island on the other. A castle appeared to stand atop the island. This castle wasn't constructed with rectangular blocks, but with meters-long columns of basalt, like the one that I had tried to lift, but far longer—and of course, far heavier. From inside the roofless structure, tall breadfruit trees rose above the height of the walls, their hand-shaped leaves seeming to wave us over from across the canal. As if on cue, a boat approached.

The boaters offered to ferry Eugene and me across the canal so we could explore Nan Madol. We spent a few hours scrambling around the structure. It was grand and mysterious, the type of ancient monument that would lend itself to paranormal hypotheses. Nan Madol is said to have inspired the settings in some of H. P. Lovecraft's fantasy novels. But, in the end, it was an empty house. To me, Nan Madol represented breadfruit's past in Micronesia—a glorious past, to be sure—but I was keen to get back to Pohnpei's present. Eugene's mind seemed to be heading in the same direction as he suggested, "let's go get *sakau*."[11]

Figure 1.1 A breadfruit tree growing behind a stone wall at Nan Madol, Pohnpei, Federated States of Micronesia. (Photograph by Russell Fielding.)

Reapproaching the house where the party was being planned, we could see that progress had been made. More people had arrived; there were now two sets of young men pounding *sakau* and the scent of cooking was everywhere. Women sat in folding chairs talking with each other and children ran around, playing. Another group of men was busy constructing an *uhm*, the Pohnpeian variation of the traditional earth-and-stone oven found throughout the Pacific. Off to the side of the *uhm*, about a dozen breadfruits lay piled, their stems tied together with a stout rope. Next to them rested a cluster of yams, wholly uprooted with the leaves still attached and clods of soil caked among the edible tubers. These were also bound with a rope and tied to the center of a long pole, suggesting they had been carried to the house atop two people's shoulders.

"You see the paramount chief has blessed the yams?" a voice from behind me asked. I turned to find a man smiling with betel nut oozing between his reddened teeth. He introduced himself as Johnny Silbanuz and explained that the party that was about to commence had been sanctioned by the region's top ceremonial leader, as evinced by the leaves-on yams displayed in the middle of the courtyard. Sometime earlier, perhaps while Eugene and I had been trekking to Nan Madol, a delegation from the family had carried the yams to the home of the region's paramount chief and asked his blessing on the event. Pohnpei consists of five separate chiefdoms, each with its own paramount chief. The island had been united when Nan Madol was constructed during a time of abundant breadfruit, but, perhaps owing to the principle that "where wealth accumulates . . . men decay," the society split soon after the breadfruit revolution, and Nan Madol was abandoned. The chief blessed the yams—and by proxy, the party—and they were carried back with the leaves still attached to attest to this fact.[12]

"Did the chief bless the breadfruit too?" I asked, pointing to the pile beside the yams. "No," Johnny said, "we can eat *mahi* anytime now since the harvest started earlier." I would learn that the first fruits of each crop harvested in Pohnpei must obtain a blessing from the paramount chief before they can be eaten, but that the blessing comes only once at the beginning of each crop's harvest season. The harvest of *mahi*, or breadfruit, had begun before I even arrived in Micronesia and the chief's blessing for the crop had long since been secured.

It turned out that the reason for the gathering was that Johnny would be leaving Pohnpei soon for several months' work in Guam. This was his good-bye party. Discussing our respective travel plans, we realized that he and I would be on the same flight. "That means it's your party too!" Johnny graciously declared.

The construction of the *uhm*—the large outdoor earth oven—was complete, its fire lit, and its stones began heating while Johnny and I talked. It was time to prepare the main course: pork and breadfruit. Three young pigs had been wandering freely around the property when Eugene and I passed by in the morning. I noticed though, when we returned from Nan Madol, that they had been tied by the ankles to stakes. Their slaughter was noisier and more prolonged than I would have liked but, finally dispatched, the pigs were gutted and splayed, then singed atop the fire. When the flames receded and only coals and heated stones remained, the pigs would be placed whole between breadfruit and banana leaves in the *uhm*.

Finally it was time to eat. By now the crowd had grown to nearly a hundred people and the sun had set. Several women, with help from the children, had plated dozens of meals and a man stood to announce the distribution of food. This man seemed to command a great deal of respect, so I leaned

Figure 1.2 A man placing halved breadfruits onto an *uhm* on Temwen Island, Pohnpei, Federated States of Micronesia. (Photograph by Russell Fielding.)

over to ask Eugene, "Is he the paramount chief?" Of course not, was Eugene's reply. You hear him talking? He's the talking chief. This made perfect sense.

The talking chief called out one person at a time, made some remarks about them, and instructed a child to bring that person a plate of food. All this was done in the local language, Pohnpeian, but when it came my time to receive a plate of food, the talking chief switched to fluent English. He thanked me for coming, and I responded that it was an honor to be included. A bright-eyed child appeared by my side with a gold-rimmed plate in her hands. "*Menlau*," I whispered, using Pohnpeian's diminutive form of "thank you." The meal was a heaping international assortment that included two kinds of rice, three kinds of noodles, a fried egg roll, pork ribs, taro root pounded into a gelatinous gray substance like Hawaiian *poi* and wrapped in taro leaves, a bunless hotdog, and an apple. I could only eat about half of what I'd been given, and then the men brought in our main course: woven palm-frond hammocks containing the *uhm*-roasted pork and breadfruit. A serving was soon placed atop my half-eaten meal, and I did my best to show my gratitude by eating the food and drinking the *sakau*.

The night ended, for Eugene and me at least, when Johnny slowly walked over to say goodnight, his eyelids nearly closed. "Going to sleep now," he told us, "I'm too *sakau*." Lewin, the early twentieth-century pharmacologist, had described the effects of *sakau* when consumed beyond moderation: "An urgent desire to lie down manifests itself. . . . The drinker succumbs to fatigue, and experiences a desire to sleep which is stronger than all other impressions." Clearly Johnny needed to get to bed. We thanked him for his hospitality, paid our respects to the talking chief, and walked to Eugene's car for the long, dark drive back to Kolonia. As we were walking, we heard the *ting-ting-ting* of stones striking a basalt platform; another round of *sakau* was being prepared.[13]

The breadfruit revolution would have involved large-scale ecological changes on any island it touched. Larger, mountainous Pacific islands may have retained some remnant of their native vegetation atop their peaks or in the depths of their valleys. Smaller or lower-lying islands could not provide such refuge. Botanists Ray Fosberg and Marie-Hélène Sachet surveyed the tiny island of Maupiti in the western Society Islands in 1985 and found "no trace of original natural vegetation" there—all had been replaced by cultivated and invasive flora. Fosberg and Sachet listed breadfruit first among the trees that "dominate the landscape." When a crop is as productive as breadfruit, the temptation is to keep planting it: if some is good isn't more better? Or, like Johnny's end-of-party lethargy caused by his becoming "too *sakau*," is it possible for a society to find itself in the condition of being "too breadfruit?"[14]

A single breadfruit can weigh up to five kilograms (eleven pounds), which is more food than a person could—or at least *should*—eat in a day. But as soon as you harvest it, it begins to deteriorate. Shelf life is not breadfruit's strength. On some islands today, where traditional foods have been largely replaced by imports, breadfruits fall to rot in the shade of their trees.

The glut of calories in the grocery store means wasted food on the ground. During times of surplus long ago, before the introduction of rice, potatoes, bread, and Spam, some Pacific Islander communities devised ways to preserve breadfruit, making a surplus last until the inevitable times of scarcity returned.

Traditionally, throughout the Pacific, the main method of breadfruit preservation has been fermentation. This is still practiced on a few islands today, including those of the Marquesas. On the island of Nuku Hiva, in a small village tucked inside the Taipivai Valley, I met Hervé Ah-Scha, a traditional maker of *popoi*, a Marquesan dish based on preserved breadfruit, fermented for several months. I had traveled to this remote part of the Pacific after leaving Micronesia to learn more about *'ulu* traditions that may be at risk of disappearance.

Hervé's outdoor kitchen resembled an old but tidy workshop. A worn wooden table served as a counter while a tied-up horse grazed beside a parted-out pickup truck nearby. Shaded by both a corrugated metal roof and a large grapefruit tree, Hervé was already at work when I arrived at 7:30 in the morning. A fire fueled by coconut husks and lumber scraps licked the bottom of a repurposed oil drum that would serve as a giant cookpot. Hervé lined the drum with large, heart-shaped hibiscus leaves called *purau*.[15]

With the fire started, Hervé added several liters of water to the drum from a garden hose, then turned his attention to the food. In a large metal bowl, grated manioc sat white and lumpy, resembling cottage cheese. Mashed and fermented breadfruit, called *ma* in Marquesan, filled a nearby five-gallon plastic bucket, where it had spent two months undergoing an anerobic transformation.

In the centuries before plastic, Marquesans traditionally made *ma* in stone-lined pits called *'ua ma*, which can still be found throughout the islands. On my way to Nuku Hiva, during a few days' stopover on the neighboring island of Hiva Oa, I had learned the old method from Félicienne Heitaa, an expert on traditional Marquesan foods. "First you dig a large hole," Félicienne instructed. "Some use a plastic bucket, but the taste is not the same." This disclaimer would later come to mind at Hervé's place. Line the hole with stones and leaves, she continued, and pick your breadfruit during *mei nui*, meaning "big breadfruit," a term referring to the period roughly overlapping the month of December, so named for the bountiful breadfruit harvest expected then. The reason for making *ma*,

Félicienne explained, is so the big harvest doesn't go to waste. This method of preservation allowed ancient Marquesans to keep some food for a future time of *mei momo*, "little breadfruit."

The next step is to peel the breadfruit and hollow out the core using a pointed wooden tool called a *mei'oka*—literally "breadfruit fork." Fill the hollowed breadfruit with seawater and let the breadfruit decompose slightly in the dark. Place decomposing breadfruits into the *'ua ma* and cover with the fronds of a coconut palm, then stones. The fermentation process can be left to its own devices for months or even years. When you uncover the pit, the *ma* is ready for consumption or for use in making *popoi*: "a good taste for old people," Félicienne's husband Gabriel added, implying that young Marquesans may have lost the taste for *popoi*.

Back in the outdoor kitchen, Hervé placed handfuls of grated manioc atop *purau* leaves, which he wrapped like burritos and gently lowered, one by one, into the metal drum's now-simmering water. When a dozen or so were in the drum, he covered them with a layer of unfolded *purau* leaves. The manioc step now complete, Hervé turned to the breadfruit. The *ma* was, as Félicienne predicted it would be, an earthy yellow-brown. It had the consistency of painter's putty and smelled bright like citrus.

From the large bucket, Hervé scooped eight large handfuls of *ma* and dropped them into another large metal bowl. He added water and, narrating the process for me in French, proceeded to *"mélange bien,"* or mix the *ma* and water well. Slowly, as Hervé kneaded and added more water in small increments, the *ma*'s consistency changed from putty to cookie dough to hummus. He quickly wrapped a dozen or so *ma* "burritos" and placed them in the drum atop the layer of flat leaves and above the cooking manioc. Now that the drum was full of *purau*-wrapped packets of manioc and *ma*, Hervé stoked the fire and told me to be patient: *"deux heures,"* two hours until it would be done cooking. As the manioc and *ma* simmered, the outdoor kitchen took on a tart fruity smell; it reminded me of a candy shop.

After *deux heures* had passed, Hervé rolled the drum off the fire and used a long-handled spoon to lift the now-cooked packets out of the simmering water. As he opened the *purau* leaf wrappings, I could see that both the *ma* and the manioc had darkened in color and their textures had become more jelly-like. Hervé retrieved a meter-long wooden board from a shelf. It resembled a small surfboard with two depressions carved into its deck—one large, one small. Along with the board, he produced a bell-shaped stone pounder, called *ke'a tuki popoi*, and prepared his work surface. He filled the smaller depression with water so it would be close at hand throughout the process and spread a layer of still-steaming manioc onto the board. Next, he spread a smaller layer of cooked *ma* atop this manioc base and wrapped it, crêpe-like, so the *ma* was fully enclosed in a manioc envelope and splashed on a little water from the reservoir to his left.

Then the pounding began. Working with a rhythm that a professional percussionist might envy, Hervé used his right hand to pound the coalescing mixture of breadfruit and manioc with the stone while his left hand alternately folded the now-forming *popoi* into layers and splashed on more water. Standing nearby, I was continuously splashed with drops of hot water and sticky *popoi*. When a batch was sufficiently mixed and pounded, Hervé divided it by hand and packaged it in small plastic bags. He sells his *popoi* at a shop in town for seven hundred francs (a little more than US$6) per bag. "*Beaucoup de travail, non?*" Hervé asked me rhetorically when he was finally done: a lot of work, isn't it? I tasted some of the finished product. It was starchy and fruitily acidic; the texture was both slippery and chewy. It was good; I'd definitely eat it again.

Whether freshly baked in a Micronesian *uhm* or fermented in a Marquesan *'ua ma*—or plastic bucket, as times change—breadfruit, prepared following the traditional recipes of the Pacific, is a hearty and tasty food. But food is only the beginning of a breadfruit tree's gifts. Nearly every part of the tree can be used for something, even the bark.

All around the Pacific, the bark of certain trees is pounded and processed to make a delicate, intricate cloth called *kapa* in Hawai'i, *tapa* in Tahiti, *siapo* in Sāmoa. Most is made from the paper mulberry tree—a species closely related to breadfruit—but the bark of breadfruit itself is also used. Older literature presented *'ulu*-based *tapa* as inferior to that made from paper mulberry, but more recent research has shown that, in several island groups throughout the Pacific, it was breadfruit that provided the material for a special variety of "high status barkcloth." This cloth was a valuable trade item throughout the Pacific islands, and, according to historian Jennifer Newell, was part of the gift package that the first English visitors received from the chiefly Pōmare clan in Tahiti: "roasted pork . . . breadfruit, plantains, and other fruits . . . and lengths of pale fine cloth."[16]

To make barkcloth, an artisan strips the inner bark from the tree and soaks it in water, sometimes for up to a month. Strips are then placed atop one another to form layers, and the cloth is softened and joined through

Figure 1.3 Hervé Ah-Scha making *popoi* on Nuku Hiva, in the Marquesas Islands of French Polynesia. (Photograph by Russell Fielding.)

repetitive beating with a heavy, fluted wooden mallet before being stamped and painted with colorful dyes derived from berry and vegetable juices. The resulting cloth finds utility in all manner of daily uses: clothing, rugs, wall coverings, or—in the case of the finest-quality barkcloth—in religious and civil ceremonies.

One cool, sunny day I walked across Central Park toward the Patricia D. Klingenstein Library of the New-York Historical Society. Months earlier, I had mentioned to another breadfruit enthusiast, art historian Billie

Lythberg, that I would be in the city, and she enthusiastically recommended making an appointment at the library. "They have a Shaw book," Lythberg said, and promised that I'd get "a real hands-on experience" if I were to go see it.[17]

Alexander Shaw was an eighteenth-century London merchant who acquired several dozen large samples of barkcloth collected during Captain Cook's Pacific voyages. With help from the naturalist Thomas Pennant, the cloths in Shaw's collection were cut down to standard sizes, compiled together with descriptive text from Cook's journal and the journals of his sailors, and published as a short run of now-rare books. Librarian Donald Kerr's exhaustive census found sixty-six copies still in existence in 2015, spread all around the world. The New York Historical, as it is now called, holds one.[18]

When I arrived at the Klingenstein Library, I showed my ID and was led into a columned, high-ceilinged space with large tables arranged in a grid. The librarian pointed me toward my table and brought out the Shaw book. I stifled a laugh when I first saw it. The book was miniscule. Maybe I was expecting something much different: a coffee-table book stuffed thick and bulging with cloth swaths, like something you would use to pick out fabric for curtains or carpet for a room. Before I even opened the book I requested a measuring tape and recorded the book's dimensions: eighteen by twenty-one centimeters, or just over seven by eight inches. It was only a couple centimeters thick.

The book's title page was covered in typescript; the official long title, in the style of eighteenth-century writing, was expansive. But what struck me were the handwritten notes interspersed upon the title page. They read, in the thick-to-thin lines of a quill pen, "presented to D Hosack by Sir Joseph Banks in 1794 and presented to the Historical Society by D Hosack New York July 1817." The book before me had been regifted to the library by a man who himself had received it from Banks, the gentleman-scientist who sailed with Cook. The regifter was David Hosack, best known as the physician who tended to Alexander Hamilton's ultimately fatal wound following his infamous duel with Aaron Burr. When Hamilton succumbed to his injury, Hosack served as one of the pallbearers at his funeral. I opened the book to a random middle page and detected the distinct odor of woodsmoke. I wondered whether this was reminiscent of Hosack, or perhaps even Banks, examining the barkcloth samples by firelight on a cool evening in New York or London.[19]

Inside the book I found an index, which listed each barkcloth sample that was to follow. Some had only basic information ("From Otaheite, used for bedding") but others came with long, narrative descriptions. One entry told of one of Cook's crewmembers going ashore at Tahiti (at the time called "Otaheite" in English) to seek provisions and coming across a group of children playing. The children, according to the narrative, surprised the sailor by surrounding him, "making many antic gestures," and one, a young teenage girl, lunged at the sailor, attempting to grab the red feathers he had stuck in his cap. The sailor, incredulous, we can imagine, took out the feathers and gave them to the girl, who quickly fled with the rest of the children. Later that day, "in the cool of the evening," according to the text, the girl—now penitent, or perhaps instructed by her parents to appear so—came to the shore where the crew had assembled, found the sailor who had given her the feathers, and "presented him the piece of cloth from which this [sample] was cut." Shaw remarked that this sample was of "the very finest of the inner core of the mulberry; and worn by the chiefs of Otaheite" and that the girl had returned to give it to the sailor in exchange for the feathers was "a true sign of gratitude in those people."[20]

Paging past the index, I came to the actual barkcloth samples. Each was cut no larger than the exact dimensions of the book itself, which made the barkcloth fragments seem like ordinary pages, each sandwiched between leaves of paper and tissue. I had expected to have to wear gloves while handling this rare book as I had during previous archival research, but the librarian told me that gloves weren't necessary, that they even contributed to damaging books since they dull the sensitive fingertips. The result was that I got to touch Shaw's barkcloth samples directly. Billie Lythberg was right: it was a hands-on experience indeed!

Each sample had its own texture, varying widely from soft cashmere to sturdy cardboard. Some were thin, slightly transparent, lacy. Others looked and felt like the patinaed cover of my mother's old, leatherbound Bible. One piece carried a note reading, "no rain will penetrate." If that's true, I would prefer a jacket made of that barkcloth to my Gore-Tex. Its texture was like felt: very soft, very smooth. Certain samples felt more like paper than cloth; indeed, some early European travelers in the Pacific had translated *kapa*, *tapa*, or *siapo* as "paper-cloth." Some samples bore their natural colors: the light tan of the breadfruit or mulberry tree's inner bark, streaked sometimes with dark blotches that resembled knots in lumber. Others had been intricately dyed in geometric patterns: sets of parallel lines, triangles, and zigzags in black, red, tan, and yellow. Some of the black-dyed pieces hid a slight bluish tinge under their aged color. I wondered if these may have been dyed with an extract of 'ulu leaves. Reminiscent of indigo, which oxidizes into the deep blue color that bears its name, breadfruit leaves, when boiled, produce a blue dye. It's tempting to imagine a single swath of barkcloth, made from the soaked and pounded inner bark of an 'ulu tree and dyed a deep blue, fading to black, from boiled 'ulu leaves—breadfruit providing both the substance and the surface.[21]

One piece stood out to me more than the rest: sample number fifteen out of thirty-nine total. It was smaller than a full page, its edges trimmed not-quite square. The cloth was very sheer and soft with a dime-size hole near the middle. It was dyed a deep reddish-brown; the color had bled over onto the paper separating this sample from the next. Rubbing the cloth between my fingers, I noted how soft and fine it felt and wondered whether it, like the piece offered to the sailor in exchange for a few red feathers, might have been "worn by the chiefs of Otaheite." Flipping back to the book's index, I found that sample number fifteen had indeed come from Tahiti, but its entry bore only one additional piece of information. I felt chills as I read: "Used at the human sacrifice."

Finally, after the gifts of fruit, bark, and leaves, there is the wood of the tree itself. Breadfruit wood is light but durable. It was used traditionally

Figure 1.4 Museum Specialist Amy Tjiong (left) and Senior Museum Technician Alex Lando (right) unrolling a large sheet of *tapa* at the American Museum of Natural History, New York. (Photograph by Russell Fielding.)

throughout the Pacific to construct small, inshore canoes used for short voyages and the components, such as the decking, outriggers, paddles, and seats, of the larger, oceangoing vessels. The very same qualities that made breadfruit such a valuable wood for these maritime uses—buoyancy, flexibility, strength, and waterproofness—lent it a special place in the history of *heʻe nalu*, the ritualized ancient Hawaiian pastime of wave-sliding, better known today as surfing.[22]

Modern surfboards are nearly always made of polyurethane foam and coated in fiberglass and epoxy. These materials were, of course, unknown to the original surfers, who carved their boards from the wood of three main species of tree, one of which was *ʻulu*. I had seen wooden surfboards hanging on the walls of surf shops from Cocoa Beach to Costa Rica, but these were clearly just for decoration. At Honolulu's Bishop Museum, a longboard made of wood stands in the corner of one display, the word *DUKE* carved into it signifying its association with Duke Kahanamoku, the famous waterman of the early twentieth century. All these examples seemed like relics. It would be easy to think of wooden surfboards in the same category of obsolescence as covered wagons or landline telephones; easy, until you meet Pōhaku Stone.[23]

Tom Stone, better known as Pōhaku, the Hawaiian translation of his surname, is a traditional board-shaper based on the east side of Oʻahu. His surfboards have been displayed in the Smithsonian and other museums and galleries around the world. But—and Pōhaku stressed this point to me—every board he makes goes into the ocean at least once, and each is perfectly ridable, even if the rest of its life will be spent hanging on a wall. That ceremonial dipping, a baptism of sorts, is part of the ritual that goes into every board Pōhaku makes. When I arrived at his outdoor workshop, Pōhaku talked me through the process.

"The best wood for me, for surfboards," Pōhaku told me, "is *ʻulu*." He lauded the wood's strength, flexibility, lightness, and natural waterproofness. "You want a board to flex" while you're riding it, he went on, so you can feel the ocean through the board, into your feet, and throughout your body. A neighbor had recently taken down a large *ʻulu* tree and Pōhaku had his eye on the trunk. He would need permission, of course, and Pōhaku knew that the neighbor was only one of the authorities to whom he must appeal. Once an arrangement had been made by which Pōhaku could take possession of the wood, a *hoʻokupu*, or ritualized offering would need to be made—not to the neighbor, who would likely be glad just to have the felled tree removed, but to the gods and ancestors that Pōhaku would need to assure that the life of the tree would continue in a new form.

What exactly that form will be, the shape of the board to come, is only partly decided by the craftsman's eye. After making a rough cut to turn a cylindrical tree trunk into a thick plank, Pōhaku offers 'awa, a Hawaiian kava beverage closely related to Pohnpeian *sakau*, to awaken the spirit of the wood so he can better see an image of the board that is to be. I couldn't help but think of Michelangelo, who famously "saw the angel in the marble," and simply carved away the excess until the angel was free. What Michelangelo was to blocks of stone, Pōhaku Stone is to planks of 'ulu.[24]

"How do these boards ride?" I asked. Pōhaku began explaining how an 'ulu board, with all its lightness and buoyancy, still floated lower in the water than a modern polyurethane board, how you must start paddling earlier to get up to speed to catch a wave, and how to steer a board without a fin underneath by actually dragging a foot in the wake. I think he could sense that I was straining my powers of imagination.

Figure 1.5 Tom "Pōhaku" Stone working on a rough-cut wooden surfboard at his home in O'ahu, Hawai'i. (Photograph by Russell Fielding.)

"Take a board," Pōhaku told me, "go surf." He warned me, though, "it's going to feel like you're trying to swim while carrying a log." With that advice, I slid a six-foot, nine-inch surfboard made from a single ʻulu plank into my rented Jeep and drove off to Waikīkī.

The board was indeed heavy; it lay low in the water as I paddled out to join the lineup waiting for a wave. Several passed underneath as I worked on the technique. Surfers on modern boards cut past me as I struggled, trying to stay on the board and simultaneously propel it forward. Without wax for grip, the deck was slippery. I needed both arms to paddle but also needed one to hold on. When I caught my first wave, though, I understood. This surfboard, unlike any I'd ridden before, felt as though it *knew* the ocean. A polyurethane surfboard is so light and rigid that it clacks and slaps with every passing ripple. It sits above the water at an almost disconnected elevation. A hard wipeout can snap one in two. This ʻulu board was heavy, smooth, and just flexible enough to bend with the rollers. It rode below the surface. Waves consumed it or propelled it forward without a sound. As the sun began to set and I decided that the next wave would be my last for the session, I let Pōhaku's surfboard carry me back to the beach silently, smoothly: *ʻulu* riding across the surface of the Pacific as it had done for generations, providing sustenance, resources, and, for me, an exhilarating ride.

CHAPTER II

Charismatic Megaflora

*A*rtocarpus altilis, as breadfruit is known by its scientific name to botanists, fits within a genus of trees in the fig and mulberry family that has diversified through natural and artificial selection into about fifty species, several of which bear edible fruits. While all are named for breadfruit—the genus name, *Artocarpus*, derives from the Greek *ártos*, bread, and *karpós*, fruit—the best-known species, worldwide, is probably the enormous jackfruit, *Artocarpus heterophyllus*, the largest of all treeborne fruits. Individual jackfruits can weigh almost fifty kilograms, or more than a hundred pounds. Botanist David Fairchild wrote during the 1940s of a time while working among jackfruit groves in a South American village when "just as I stepped under the tree there came at me a falling object that seemed as large as the side of a house." He quickly jumped out of the way of the falling fruit and, "with a splash there lay on the ground before me a mass of green rind and golden yellow custard that covered four feet of the dooryard." The fruits of other *Artocarpus* species are not as big, but most do share the rind-and-custard structure of the jackfruit. Within the genus, alongside jackfruit and its other relatives, is what botanists call a "breadfruit complex," comprising three closely related tree species, only one of which is the true breadfruit.[1]

The first species in the breadfruit complex is believed to be the ancestor of the other two. Commonly called "breadnut," its scientific name is *Artocarpus camansi*. The fruit of the breadnut tree is bright green and oblong, about the size of a cantaloupe, its thick rind covered with soft spikes. Each mature fruit contains fifty or more seeds, soft and cream-colored when the fruit is still on the tree, that transform quickly after harvesting into hard brown nuts. These nuts are the main food product derived from the tree, as the small amount of flesh produced by the fruit is highly sensitive and deteriorates quickly. As a traditional food of New Guinea and the Philippines, breadnuts are eaten in stews or salads; they can also be boiled or roasted, and they resemble chestnuts in both flavor and texture. This resemblance is so strong that when the Spanish colonized the Philippines, they called breadnut *castaña*, the same name given to chestnuts.[2]

The second species in the breadfruit complex is *Artocarpus mariannensis*, commonly called "dugdug." This fruit tree grows natively only in the Mariana Islands, Guam, and Palau and has been transplanted to a few other island groups in Oceania. The dugdug is a funny little fruit growing on a massive, stately tree. These seed-bearing fruits look something like a breadfruit that's been deflated or, well, sat upon. The trees are adapted to the low, coral atolls of their native region and tolerate higher levels of salinity than either breadfruit or breadnut. They stand tall, supported by huge buttresses at the base of the trunk. On the island of Saipan I saw a large dugdug tree growing in a family's yard. Its buttresses had engulfed a small, wooden workbench. Rather than cut down the tree or even cut away some of the trunk, the family had simply removed several planks from the bench's tabletop and allowed the tree to keep growing. Fruit bats love to eat dugdugs, and they disperse the seeds freely through their droppings.[3]

Third and finally, we have the true breadfruit, *Artocarpus altilis*, 'ulu. A mature 'ulu tree grows tall and wide, casting a shade most welcome in the tropical sun. If left unpruned, the branches—easily climbable—begin low and twist beneath the foliage, straining upward while remaining well anchored to the tree's thick, knobbed trunk, extending far higher than the longest picking pole can reach. For this reason, most cultivated trees are pruned in such a way as to limit upward growth and to encourage lateral spreading.

The trunk of an old breadfruit tree is covered with splotchy, brownish-gray bark featuring sags and cracks not unlike a mature elephant's skin. Younger trees, and younger branches on older trees, have smoother bark. The knots left by branches that have broken off testify to the soft wood's susceptibility to strong winds. Lower, older branches can start as thick as the trunk itself. Newer branches, higher on the tree, are segmented and

Figure 2.1 A breadfruit tree in Hawai'i (left); a breadfruit leaf in Bermuda (center); and a breadfruit in Barbados (right). (Photographs by Diane Fielding, left, and Russell Fielding, center and right.)

serpentine, resembling gray, ringed snakes, often with scattered light-gray circular markings and tiny reddish hairs. Thin branches bend easily and spring back into position when released as though they were elasticized. The bark's surface is often incised like the face of a Maasai hunter. These scars are not all the work of sap collectors or graffitists; they go all the way up the trunk and out onto the thin branches—well beyond the reach of any blade. Some of these growth scars are like little habitats of their own, supporting tiny algae, sheltered within the incisions. Although carpenters do not consider breadfruit a hardwood, the trunk of the living tree feels very solid. Knocking on a breadfruit tree as though it were a neighbor's front door produces no echo. Pulling yourself up onto a thick, low branch has no discernable effect on the tree; it can support a human climber as well as it can the multitude of bright green anoles or geckos that scurry up and down the highways of the trunk and branches.

The leaves are broad and bright green on top, lighter gray-green on the underside. Though the exact shape of a breadfruit tree's leaves depends upon the specific variety, most, in the words of author Jonathan Drori, are "deeply lobed like Matisse stencils." Their silhouettes are unmistakable: giant, multifingered hands undulating softly in the breeze, alternately protecting and displaying the fruits that hang beneath, singly or in small clusters, from fleshy thick stems. If the breeze is gentle, a soft rustling sound is heard; stronger wind, especially accompanied by hard rain, prompts a crashing din from the great surface area of the leaves. You'll hear the same sound, the crashing, when large birds or other arboreal animals move among the branches of a breadfruit tree.[4]

New leaves unfurl from conical growths at the tips or along the lengths of the branches. The newer leaves feel waxy. Older ones have attained a rougher texture. Running your fingers along the underside of a mature breadfruit leaf feels like stroking a sheet of 200-grit sandpaper. If the site where a tree grows is exposed frequently to strong winds, many of the leaves will be torn and some branch tips will have no leaves at all. If you crumple a fresh green breadfruit leaf, it stays crumpled in your hand without springing back. The crushed leaf tastes astringent. At the base of the tree, unless someone has been by with a rake, a pile of fallen leaves litters the ground, most having lost their chlorophyll and now displaying a deep tan color and offering a crunch to the step.

Growing up in the rural outskirts of Tampa, Florida, one of my regular and least favorite chores was to pick up the Spanish moss that had fallen from the many live oak trees on our property. On Moʻorea—Tahiti's smaller (and lovelier) neighboring island—I met Hinano Teavai-Murphy, who described the main domestic task of her childhood: picking up the leaves and other droppings of the family's large, front-yard breadfruit tree. I sympathized. But Hinano gained something from her experience that I did not get from mine. Throughout the year, she noticed that she was picking up different parts of the tree during different seasons: dried leaves, fallen flowers, broken branch tips, ripe and rotting fruits. "You learn the cycle," she explained to me, by noticing what falls from the tree at different times of the year. You learn to anticipate breadfruit season, and to be ready when the Pleiades constellation, called Matariʻi in this part of French Polynesia, rises again to signal the beginning of the harvest.

Emerging here and there from the ends of branches, male and female flowers appear on short stalks, cradled within the protective cover of their leaves. A single tree produces flowers of both sexes. Male flowers are long, yellow catkins vaguely reminiscent of bent, fuzzy candles. After producing pollen, the male flowers fall to the ground where they turn brown and look like overripe bananas. Picking one up, you feel like you're holding a cloth pencil case. Rubbing the surface with only slight pressure removes the velvety outer layer, revealing a slender, slippery core. Female flowers look like small breadfruits, not yet developed. Both female and male flowers are called "inflorescences," meaning they are not single flowers but agglomerations of thousands of tiny flowers clustered around spongy cores. While only the female flowers develop into fruits, male flowers can be dried and burned to repel mosquitos and other insects. They can be candied and eaten. Hinano told me that as a schoolchild, she used male breadfruit flowers to erase the chalk on her writing slates during lessons. Hinano's favorite way to prepare breadfruit is to roast it slowly over a fire fueled by coconut husk and 'aito, or Pacific ironwood, then scoop out the fruit's core and fill the hole with coconut milk. When the flesh of the breadfruit absorbs the milk of the coconut, it's ready to eat—a delectable treat made from the two most important tree crops in the Pacific.

Breadfruits, the fruits themselves, resemble bright green cantaloupes from afar. The smallest, youngest fruits stand up atop their stems like lollipops. As the fruits grow with age, they lean downward to hang below. Younger fruits are very solid and dotted with exuded drops of white sap. Larger, ripening fruits feel a little softer to the squeeze. Most are roughly spherical, though there are some varieties that produce oblong fruits—more the shape of a mango than an orange. The surface color ranges from fluorescent green to reddish brown.

A breadfruit's skin is covered with small, rounded bumps. In some varieties, these bumps rise into small spikes; in others, they lie almost flat, forming a nearly smooth skin. Some varieties change from spiky to smooth as the fruit ripens. Closing your eyes and rubbing your hands over a breadfruit, you might imagine that you touch something not vegetal but reptilian. "Dragon eggs" is a common comparison, both intentionally and accidentally referencing *Game of Thrones*, in which author George R. R. Martin described "the egg with the deep green shell, bronze flecks shining amid its scales." Etched onto the skin of a breadfruit is a matrix of irregular but roughly hexagonal shapes that indeed resemble scales, each with a small dot or point roughly centered. Breadfruit skin looks like what you see when viewing plant cells through a microscope.[5]

The "cells" or polygons etching the skin of a breadfruit are the outermost surfaces of the thousands of flowers that combine to form a single fruit. The fruits of most flowering plant species grow as the matured ovaries of single flowers: each apple grows from one apple blossom, each orange from a single orange blossom. Similar to a pineapple, though, a single breadfruit grows from the coalescence of many flowers. The two thousand or so individual flowers that comprise a single breadfruit all grow outward from the fruit's core.[6]

The perianths, or outer parts of each flower, continue to swell and become fleshy, forming the edible part of the fruit. When breadfruit is harvested at the mature stage, the most common stage for consumption, this flesh is white, sometimes tinged with yellow, but begins to gray quickly after you cut it with a kitchen knife, exposing the interior to the atmosphere's ambient oxygen. The texture of the flesh depends largely upon the ripeness of the individual fruit: less ripe breadfruit is hard and dry; riper fruit is softer, moister, and beginning to attain a sweet smell. Most ʻulu varieties originating from the central and eastern Pacific yield seedless fruit. Those that do seed, like the ʻUlu fiti' variety from Fiji, produce seeds spaced irregularly throughout the flesh. While the seeds cannot survive drying and are only rarely used for propagation, they can be cooked in the fruit or removed and roasted separately. They taste like chestnuts and are high in protein. In the Pacific, the presence of seeded breadfruit varieties diminishes as you move west to east. Since the Pacific islands were settled from west to east, this geographical progression of increasing seedlessness indicates a preference for seedless varieties among the original dispersers, who, preparing each time for their next voyage, would have carefully selected individual trees from which to collect root shoots for propagation on islands over the eastern horizon.[7]

Each fruit is attached to the tree via a thumb-thick cordlike stem; picking breadfruit takes effort! Unpruned, an ʻulu tree grows tall and produces fruit all the way to the top. The cultures of nearly all Pacific islands where ʻulu grows have produced specialized tools for their picking: usually consisting of long poles with forked or hooked ends, nets or baskets lashed beneath, to sever the stem and then gently catch the fruit so it can be lowered, unbruised, to the ground.

Even the longest picking pole cannot reach the tantalizing topmost breadfruit, which remains on the tree until it falls naturally, or is climbed to by a brave—or desperately hungry—picker. The motif of the high-growing breadfruit is common in Hawaiian proverbs, with admonitions to take the ʻulu i ke alo, the "breadfruit just in front," rather than strain for one out of reach. This is meant as advice to look nearby, both in politics and in love, choosing a loyal peer, someone just in front, as a political ally or a romantic partner, rather than ka ʻulu loaʻa ʻole i ka lou ʻia, "the breadfruit that even a pole cannot reach," as is said of an untouchable person of very high status.[8]

When you do manage to break the ʻulu off its stem, you release a trickle of milk-white sap—more than a trickle if the fruit you've picked is unripe

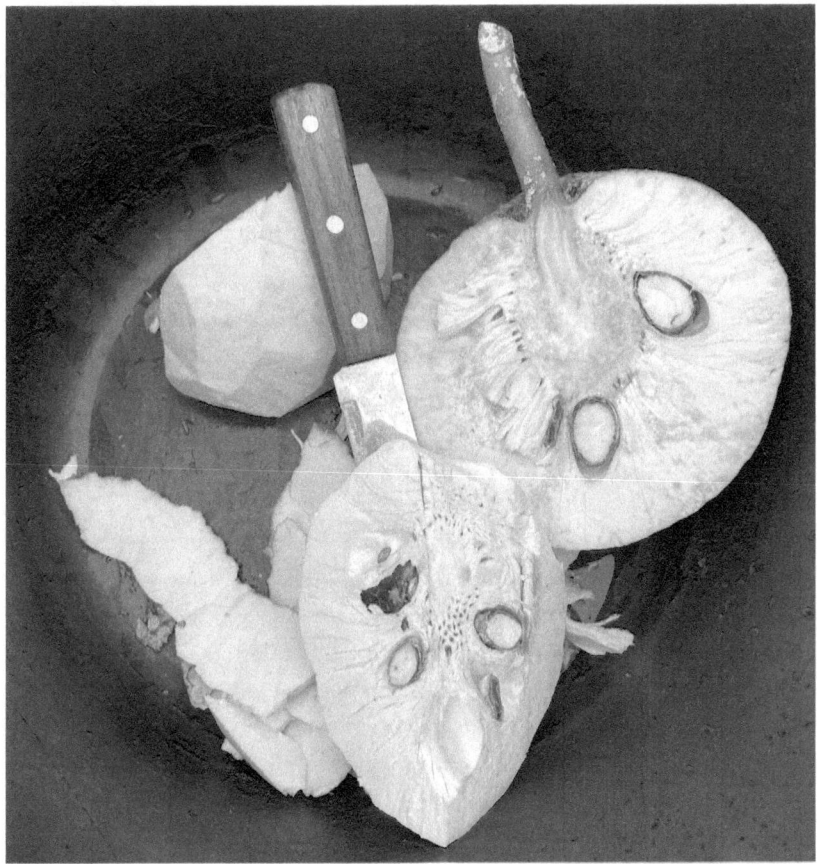

Figure 2.2 A sliced breadfruit with seeds in Ghana. (Photograph by Russell Fielding.)

or if the tide is high. Like seawater, breadfruit sap responds to the moon's gravity, moving upward, into the branches and fruits during high tides and settling lower in the trunk when the tide is low. The sap looks and feels like Elmer's glue, an opaque white, sticky, and only slightly viscous liquid. It smells like freshly cut grass and tastes like warm oat milk. Left to dry into a latex, its viscosity and adhesiveness increase.

This sticky latex has found many traditional uses, ranging from sealing gaps in the hulls of voyaging canoes to providing an antidote to ciguatera fish poisoning, a scourge to communities reliant upon coral reef fisheries for their subsistence. Traditional practices throughout the Pacific include methods of trapping birds in this sticky sap, a process called "bird liming."

In Hawai'i, bird liming was done mainly to collect the red and yellow feathers used to make the cloaks and helmets worn by the *ali'i*, or divine chiefs. After a trapped bird had been plucked of small amount of plumage, the trapper would dissolve the breadfruit sap using the oil of the *kukui*, or candlenut. The bird was then free to fly away, regrow its feathers, and, perhaps, if it hadn't learned its lesson, to be trapped again another day.[9]

Holding the breadfruit in your palm, fingers splayed for support, consider it up close. It resembles a green globe, the severed stem at the north pole and its antipode resting heavily in your hand. It's dense like a melon. Halving the breadfruit from pole to pole best reveals the botanical structure: a broad, rounded core continuing from the stem downward to a little past the midpoint of the fruit. The thousands of fused flowers that make up the fruit's flesh, each radiating outward, are attached to this core at their interior ends and to the rim of green skin at the outer. The fusing of the individual fruited flowers is not entirely complete; elongated gaps are visible, usually closer to the core than to the skin, which lend a slight spongy texture to the bite when you eat breadfruit. These gaps appear as clusters of small holes in the flesh of a breadfruit sliced through its equator.

The taste of *'ulu* depends largely upon when it was picked and how it's prepared. Usually harvested at its "mature" stage—that is, ripe but not *too* ripe—the flesh of *'ulu* is starchy and firm. Fully ripened, or overripened, breadfruit develops a profuse sweetness that can be captured in a range of delicious desserts. Young, unripe *'ulu* is usually left on the tree to mature, but if picked too early or blown down by the wind, immature *'ulu* can be pickled. In taste it resembles artichoke hearts.

Breadfruit is usually not eaten raw, except when in ripeness it takes on a syrupy sweetness and pudding-like texture. Even cooked, when presented alone and unseasoned, it can be offputtingly bland. One of the simplest ways to prepare *'ulu* is to roast it whole atop a fire fueled by wood, charcoal, or coconut husks. Wait until the skin turns from green to black and turn the breadfruit occasionally to ensure that its entire surface is scorched. Then remove it from the fire and scrape the charred skin away to reveal the perfectly roasted flesh inside. And here the inevitable comparison to potatoes will begin.

I've met one farmer who refers to *'ulu* as "tree potato," although I think a blindfolded taster would have no trouble distinguishing the two. Roasted *'ulu* tastes starchy and smoky. Baked or roasted potatoes have a grainy texture entirely absent from roast breadfruit, which seems to bounce back ever slightly

Figure 2.3 A halved breadfruit. (Photograph by Becky Hadeed, The Storied Recipe.)

under the pressure of the bite. Masticated potato reduces to something finely granular, but when you chew breadfruit the texture produced is silky. Any seasoning you might enjoy on a baked potato will enhance the taste of your roasted *'ulu*: salt, pepper, butter, or garlic would all be good starts.

A trip to the produce section of your local grocery store may yield fresh avocados from Mexico, bananas from Ecuador, oranges from Florida, and pineapples from Costa Rica, but if you don't live where breadfruit is grown, probably no *'ulu*. Breadfruit deteriorates soon after picking, making it difficult for growers on isolated and remote islands like those of the Hawaiian archipelago or, for that matter, most of the rest of Oceania, to ship the crop long distances. Refrigerated shipping containers can prolong the time to spoilage, and new postharvest technologies are currently being introduced, but until recently breadfruit has been a largely *im*moveable feast.

Freezing is not kind to the taste and texture of whole, raw *'ulu*, but when it's been peeled and lightly steamed first, frozen breadfruit can retain its quality for up to a year. Fried *'ulu* chips, or crisps, are gaining popularity as a shelf-stable and easily transportable way for growers to supply markets both near and far. Breadfruit chips tend to be cut thicker than typical potato chips and as such have a slightly more resistant crunch. Perhaps owing to this thickness, breadfruit tends to absorb a noticeable amount of the oil in which it's been fried, which imparts a slight canola, coconut, palm, or peanut flavor to the chips. In other words, you should choose your frying oil carefully.

One of the most popular modern ways to convert *'ulu* to something shelf-stable is to mill it into flour. With the stem removed, an *'ulu* placed upside down will drain most of its sap in a few minutes. You can then peel the fruit if you want your flour more uniformly white or leave the skin in place for more of a "whole wheat" appearance and taste. After draining, the fruit is quartered, then sliced or shredded, before drying in an oven or in the sun. Dried breadfruit is ground into flour, which can retain its quality for at least a year before going stale. Most bakers use a mix of breadfruit and wheat flours in their recipes but for fully gluten-free cooking, breadfruit flour can be used alone—just don't expect the same lift or texture as when baking with wheat flour.[10]

Often, from the ground under the shade of a breadfruit tree, new young trees will sprout directly from the roots. Botanists call these offspring "root suckers" or "root shoots;" Hawaiians call them *keiki*, the same word they use for their own children. (In both Bermuda and Florida, I heard them called "pups.") These "children" are clones of the parent tree, genetically identical. By propagating a breadfruit tree from its root shoots, you can be sure to get a tree with the same desirable characteristics as its parent: broad canopy, reliably seasonal fruiting, seedlessness. Even breadfruit varieties that do produce seeds are not propagated by planting those seeds because the seed-grown breadfruit trees often are not "true to type," meaning they may not exhibit the same characteristics as the parent plant. Wild ancestral breadfruit trees, most of which produce seeds, expand into diverse groves with a range of characteristics seen in the trees, leaves, and fruit. As domesticated breadfruit developed through human selection to produce fewer and fewer seeds, people recognized the consistency of propagation through root shoots and began to use that method to ensure future generations remained mostly seedless.[11]

Breadfruit trees are still propagated in this way today, since their original distribution throughout the Pacific, with very little technological change beyond the replacement of leaf wrappings with plastic buckets and

Figure 2.4 A breadfruit root shoot (left), growing from its parent tree (right) near Pā'ea, Tahiti. (Photograph by Russell Fielding.)

the addition of synthetic fertilizer. Soil is removed from above roots that have sprouted *keiki*, a short section of root is cut out with the shoot attached above and some soil below, and this section is replanted in a propagating bed or pot. This is a crucial time in the growth of a breadfruit tree. The Breadfruit Institute, established in 2003 as part of the National Tropical Botanical Garden in Hawai'i "to promote the conservation, study, and use of breadfruit for food and reforestation," advises that the type of soil, level of moisture, and amount of shade are all critical factors. If nurtured well, the cut root shoots will grow into small breadfruit trees about a meter tall, which is the height at which they're ready to be prepared for field planting, ideally at the onset of the rainy season. It takes about two months of gradual introduction to full sun before young breadfruit trees are ready to begin photosynthesizing to their full potential; remember, they were "born" under the shade of their parent tree's canopy. After this period of acclimatization, the lower leaves on the small trees are cut to reduce the amount of water lost to transpiration and the trees are planted in the earth. According to Diane Ragone, if you follow these instructions closely, "success rates close to 100% can be expected."[12]

While propagation by root shoots works extraordinarily well, faster ways exist to cultivate breadfruit. One method, which, to the novice horticulturist might seem like sorcery, goes by several names including "air-layering" or "marcottage." In Florida, it's known as "mossing-off," owing to the importance of sphagnum moss to the process. Whatever you call it, this method of propagation involves inducing a plant to produce structures known as "adventitious roots," roots that occur not underground but at the base of a branch, so the branch can then be removed from the tree and planted on its own.[13]

Air-layering, then, may be thought of as the opposite of grafting which, too, was once seen as a form of magic, considered "a secret by the initiated and a miracle by the public," according to fruit historian Frederic Janson. The first step in air-layering, I'm sorry to say, is to injure the tree. Find an upward-growing branch that will become the stem of the new tree. Carefully remove the bark in a strip all the way around the branch, slightly above the place where it joins the main limb. This girdling interrupts the downward movement of carbohydrates created by photosynthesis in the leaves, causing this sugary plant food to accumulate at the site of injury. Growth-promoting hormones called auxins, which are produced at the tip of a growing branch, also accumulate here naturally or can be added by the person doing the air-layering. Now, with carbohydrates for fuel and hormones to initiate root growth all that's needed is a medium into which the adventitious roots can grow. This is accomplished by surrounding the girdled section of branch with organic matter—usually some mix of moist soil, mulch, or moss—and encasing the mass with a cloth or plastic enclosure. Cover the entire assemblage with something opaque to keep out the light. No need to be overly technical here; horticulturists often use aluminum foil. Roots grow in the dark, after all. Then you wait for the sorcery to happen.[14]

When the adventitious roots have grown, carefully cut the branch and remove it from the parent tree. This new young tree should be planted in a pot and kept shaded; too much of anything—handling, sunlight, even noise, it seems—can be lethal. Once you've observed signs of growth and you're confident that the new tree is thriving, it can be planted in the ground beside its parent tree or halfway around the world. Trees propagated by air-layering usually produce fruit sooner than those grown from root-sprouted *keiki*.[15]

Another modern method of breadfruit reproduction is through tissue culture. At laboratories around the world, cells of specifically selected breadfruit trees are cloned into whole new breadfruit plants. These test-tube *keiki* can be mass-produced and shipped long distances quickly, allowing for the rapid planting of entire breadfruit groves. I'm not the only one, though, who worries about the implications of mass-cloning 'ulu. The history of another domesticated fruit suggests the need for caution.

During the mid-twentieth century, a fungus known as Panama disease all but destroyed the commercial banana industry in Latin America. Nearly all plantations then were growing clones of a variety known as 'Gros Michel,' sometimes anglicized as 'Big Mike,' which was susceptible to the fungus. Banana growers had to replace their crops with a new variety resistant to Panama disease and, perhaps unwisely, most switched to clones of 'Cavendish,' the variety of banana found in almost every North American supermarket today. The same lack of diversity, though, that led to the loss of 'Gros Michel' may very well cause the demise of 'Cavendish' if something like Panama disease emerges to target this newly widespread variety. Similarly, if a blight ever attacks one of the few commonly cloned breadfruit varieties, many growers will lose their trees.[16]

In Tahiti I met a breadfruit grower who has learned the lessons of the past and is planning for a more diverse, more resilient future. Pevatunoa Levy lives quite literally at the end of the road. The island of Tahiti is shaped like a figure-eight with a large, oval island, Tahiti-nui ("big Tahiti"), joined to a smaller, more circular island, Tahiti-iti ("little Tahiti") by the narrow Isthmus of Taravao. Tahiti-nui is where the capital, Pape'ete, is located. It's home to the international airport at Fa'a'ā and almost 90 percent of the island's population. Tahiti-nui is naturally beautiful, but there's a lot of traffic, a lot of concrete. Tahiti-iti is more distant, laid-back, and probably best known for the "ultimate wave," a perfectly formed, reliable surf break just offshore from the village of Teahupo'o and the site of the surfing competition in the 2024 Olympic Games. While athletes from every other Olympic sport competed halfway around the world in France, the surfers were in the South Pacific, each hoping to catch an ultimate wave in front of the judges.[17]

I drove my small rental car down to Tahiti-iti to meet Peva, as he's called, who had been described to me in Pape'ete as a breadfruit *grower*. What I found instead was an enthusiastic breadfruit *collector*. Through family inheritance, Peva has come to own much of the land in a long, lovely valley that stretches inland from the Teahupo'o shoreline toward the peak of Mount Rooniu. Peva, barefoot, asked me if I'd like to go for a walk. About a mile into the hike, which took us over grassy fields and through smooth-rocked streams, I decided he had the right idea, and I kicked off my sandals. As we walked, Peva pointed out individual breadfruit trees and shared some quick facts about each variety: 'Piri'ati,' two fruits grow together, one big, one small; 'Havana,' his favorite; 'Pua'a'—here I interjected: "Pig!" There are not many words I've learned in Tahitian beyond the obligatory hello (*ia ora na*) and thank you (*māuruuruu*), but *pua'a*, meaning "pig," is one of them. "Do you feed this one to pigs?" I asked. No, Peva told me, but it's good to eat with pork!

We passed a fenced paddock with several horses grazing. A waterfall in the distance marked the head of the valley. Peva's knowledge of breadfruit varieties was encyclopedic. Most of those he knew, he also grew here in his valley. Some of the rarer ones had taken considerable effort to locate, to propagate, and to nurture into fruiting. "Why do you do it?" I asked. Peva explained that he started his collection to preserve the breadfruit diversity he had seen within the Pacific. Varieties like 'Ma'afala' were

Figure 2.5 Pevatunoa Levy, walking on his property near Teahupo'o, Tahiti. (Photograph by Russell Fielding.)

becoming dominant, and the rarer ones were disappearing. "If we don't do this, we'll lose it," he told me, referring to the amazing diversity his ancestors had created through careful selection, "Who will do it? I will!"

Thanks to experts like Peva who have the knowledge of breadfruit varieties, the understanding of why botanical diversity matters, and—crucially—the land on which to grow trees, the gifts of ʻulu will continue to be enjoyed by another generation of Pacific Islanders. Will anyone else carry on after Peva is gone? He hopes someone younger than himself will step up and say, "I will."

CHAPTER III

The Treeness of Life

The settling of the Pacific was fueled by *'ulu* and the trees growing throughout the region today help us understand how that settlement took place. Almost forty thousand years ago, small bands of hunter-gatherers left the large island we now call New Guinea in bark canoes, dugouts, and rafts moving east. They found, relatively close offshore, the archipelagos now known as the Bismarck Islands, which today are part of Papua New Guinea, and the Solomon Islands, now an independent country. Moving east, these explorers continued to find one pristine island after the last. Theirs were the first human footprints ever to mark these sands. But when they reached small Makira Island, they stopped their eastward movement. Why?

Probably because they lacked the boats and navigational skills to travel any further. Off the east coast of New Guinea, islands in the Bismarck and Solomon archipelagos are "intervisible," meaning that, before you lose sight of the island you're sailing from, you can see the next island rising over the horizon. The standards, then, for seaworthiness of vessels and skills of navigation, are low when sailing between intervisible islands. It would have been possible simply to paddle a raft or canoe during calm weather toward an island whose peak was visible on the horizon. Past Makira, however, there's a lot of ocean to cross before you see, let alone reach, the next island.[1]

By about ten thousand years ago these people, whom archaeologists call Papuans, had settled down in what, to them, must have seemed like the end of the world, the last islands in a long archipelago. Life was good. The Papuans had developed resistance to malaria, which was prevalent in the area. New forms of agriculture were emerging as they began to domesticate wild food crops, including bananas, taro, and yams. It was also here, in the islands off the east coast of New Guinea, and possibly on New Guinea itself, that breadfruit was first domesticated from its wild ancestors.

After originating on the island of Borneo, the botanical ancestors of breadfruit are thought to have been first distributed throughout Southeast Asia and beyond by animals like elephants, flying foxes, and possibly some species of megafauna that are now extinct. While some tree species in the *Artocarpus* genus have been domesticated and are mostly propagated by human initiative, others remain wild and reliant upon natural methods to maintain their habitats. Conservation biologist Evelyn Williams has warned that, "as the range of the modern Asian elephant and other large mammals shrinks, so too may the dispersal of *Artocarpus* species."[2]

Wanting to see for myself the environment in which breadfruit's ancestors had evolved, I traveled to Borneo to learn more about the wild fruits still growing there and how the local population uses them. Borneo is a large island shared by three countries: Brunei, Indonesia, and Malaysia. After arriving in the Malaysian city of Kuching, I met my guide, James Chee, and we set off to explore the markets and jungles of his home island. At produce markets spread throughout the state of Sarawak, James and I saw wild fruits collected from the jungle—mainly by members of the Bidayuh and Iban cultures—presented alongside domesticated crops grown locally on Bornean farms and fruits and vegetables that were more familiar to me, imported from overseas.

We saw *buah entawak*, a wild relative of breadfruit about the size of a navel orange and covered with soft, bristly spikes. Inside are capsules of reddish flesh that taste like pumpkin. We saw *buah kulim*, known colloquially as jungle garlic. The word *buah* simply means "fruit," which is why it appears so frequently in the names of Malaysian produce. Kulim fruits look like golf ball–sized coconuts and produce an oil that is used both to season food and as a medicine. We saw marang, another relative of breadfruit that, when opened, reveals dozens of small, white fruit pods—technically called "arils"—that look like marshmallows and taste like pears. Tropical fruit expert and guide Lindsay Gasik has described marang as being "so sweet it's like someone made a cake juicy." Trekking into the forest near the village of Siburan with one of James's friends named Beko, we saw trees bearing *buah kepayang*, a large, brown fruit about the size of a breadfruit but slightly more oblong, like a mango. Beko explained that if you eat *kepayang* fresh, you could become intoxicated or worse—you could die from the poison. Soaking the fruit for up to a week removes the toxin and renders it safe for consumption. Beko's father had been a shaman and used the oil of the *kepayang* fruit in his spiritual practice.[3]

At the Kubah Ria market, a vendor stopped me to ask where I had come from. "United States," I told him. "Ah, the U.S.," the vendor replied, smiling and gesturing toward his wares, which included *jering*, an edible seed collected from a wild tree in the Bornean forest. "Plenty high technology there, but you don't have this!" Clearly the abundance that exists among the islands on which breadfruit evolved and was first domesticated provides an enviably good life in the eyes of the people who live there.

But the good life the Papuans had created for themselves wasn't the only reason they stopped migrating. They stayed where they were also because there was no place else they could go. Moving west would only bring them back to the islands whence their ancestors had come. North was Asia; south was Australia—both already inhabited. And east? As far as they knew there was nothing but seawater, forever, to the east. So, for thousands of years, the Papuans were a settled people, living and farming in New Guinea and its nearby islands. But some of these Papuans may have felt a deep—genetically deep—sense of restlessness.

Every person, in their DNA, has a gene that scientists call DRD4. The DR stands for "dopamine receptor," which means that this gene is associated with dopamine, a hormone linked to the brain's reward system and to feelings of pleasure and satisfaction. As is true for other genes, several variants, called "alleles," of DRD4 are found within any given population. For example, different alleles of the genes responsible for eye color give us the blues, browns, greens, and all the in-between shades that we see in the eyes of our friends and relatives.[4]

Since DRD4 seems to affect the development of one's personality, researchers have been interested in learning whether each of this gene's alleles might be associated with particular traits, shared by people who share that allele. And because of DRD4's association with feelings of reward and satisfaction, might it be the case that the brains of people with different DRD4 alleles feel satisfaction from different rewards?[5]

It seems so. Two alleles of the DRD4 gene, called 7R and 2R, together are found in about thirty percent of the world's population and these alleles

Figure 3.1 Fruits, including *marang* (foreground) at the Serian Daily Market in Sarawak on Borneo, Malaysia. (Photograph by Russell Fielding.)

appear to be associated with an embrace of movement, change, and adventure. People with 7R or 2R in their DNA often exhibit more curiosity and less risk-aversion than do those with other alleles. They travel more and they take greater risks. They're more restless. A 1996 study led by geneticist Fong-Ming Chang showed that this effect can be seen at the level of the society, not just the individual. Chang found that these alleles, which have come to be called "explorers' genes," are more common in cultural groups with migratory histories than in sedentary ones; the collective restlessness seems to have influenced the movement. It was no surprise, then, to find that modern Papuans, whose ancestors had migrated to the very edge of the accessible world, have a greater concentration of these explorers' genes than nearly any other ethnic group of Eurasia.[6]

Fortunately for the ancient Papuans, their restlessness took them to a place of abundance. But even though life was good on the tropical shores of the Bismarck and Solomon archipelagos, with their abundant seas and productive breadfruit groves, some may have been driven to keep moving. You can just picture them, these restless adventurers, the need to explore encoded in their DNA, standing with their toes at the waterline on the east-facing beach of a tiny island, yearning to go further, to see what lay over the far horizon. You can imagine them cursing the insufficiency of their watercraft that were prone to flooding or breaking apart in heavy seas.

And then, the Austronesians arrived. About 3,500 years ago, a new group moved south from eastern China and Taiwan into what is now Indonesia, across New Guinea, and into the eastern archipelagos inhabited by those restless Papuans. The Austronesians moved quickly, using large sail-propelled outrigger canoes and other new maritime technologies not previously seen in these islands. Here, the term *canoes* should not be thought of as anything resembling the small watercraft that's a piece of standard equipment for summer camps everywhere. Pacific voyaging canoes, or *vaka moana*, are massive ships, capable of carrying entire extended families and everything they would need to establish a life on a new island. The Austronesians' confidence in navigating their canoes came from generations of using them to cross back and forth between Taiwan and mainland Asia and to travel along the Asian coast.[7]

When the Austronesians sailed into the Bismarck and Solomon archipelagos, they did not so much replace the Papuan way of life as *enhance* it. The two groups merged peacefully, and in doing so, they formed a new culture with strengths inherited from both sets of ancestors. The Austronesians brought new technology like their canoes and new skills like the ability to navigate using the stars and other environmental clues. Navigators would memorize the rising and setting locations of hundreds of stars—these points on the horizon were called "star houses"—and their positions could serve to orient a navigator on the open ocean, far from the sight of land.[8]

The Austronesians could fish in deep water from the canoes using technologies the Papuans had not yet invented, and they introduced new art forms like pottery and tattooing. The Papuans brought knowledge of tuber and tree-crop agriculture, like breadfruit, which was more productive in these equatorial latitudes than the rice the Austronesians had known in Asia. The Papuans also contributed malaria resistance, and, it seems, the restlessness and yearning for adventure encoded into their genome. For all their travel-related skills and technology, the Austronesians did not have especially high concentrations of the explorers' genes. In other words, the Austronesians *could* travel; the Papuans *needed* to travel.

The new culture they formed is known as the Lapita people, named after an archaeological site in New Caledonia where some of the first physical evidence for its existence was found. The Lapita had the tools and skills required for oceanic exploration and the driving urge to set forth and sail to the east. But sailing eastward into the unknown, with no land visible on the horizon, was no easy matter. These sailors needed to hone their skills before taking on open-water crossings longer than any their ancestors had seen before. Fortunately, this culture arose in an area that anthropologists would later call a "voyaging nursery," owing to its intervisible islands, reliable winds, and abundant resources. The Lapita would stay in these islands to the east of New Guinea for hundreds of years, developing into the greatest voyagers and navigators the ancient world had ever seen.[9]

When they were ready, they went. About three thousand years ago, Lapita expeditions began to leave the Bismarck and Solomon archipelagos sailing eastward. Canoes were filled with family members, domesticated dogs and pigs, stone tools made from the volcanic rocks of the Bismarck Islands, shell fishhooks, adzes made from the shells of giant clams, and the seeds and root cuttings of crops that would be needed to establish familiar agricultural systems on the new islands they would discover. The breadfruit and other crops the Lapita distributed throughout the Pacific during their expansion have helped archaeologists piece together the precise path of their migration. In the Chuuk Islands, the linear cluster of stars spanning across the night sky that we now call the Milky Way was named "Anenimey," which means "path of the breadfruit." Voyagers bringing ʻulu in the hulls of their canoes were also following constellations of ʻulu in the sky. Finally, these deeply restless explorers were on the move.[10]

The Lapita first arrived in the Santa Cruz Islands, which today comprise the easternmost part of the Solomon Islands but were until then completely unknown to humans. They kept going at a quickening pace. The Vanuatu archipelago and New Caledonia were found and settled just a few hundred years after the voyaging commenced. Fiji soon after, then Tonga and Sāmoa. According to Pacific anthropologist Patrick Kirch, it took the Lapita "a mere 250 years" to expand their world one-third of the way across the Pacific Ocean. These were not one-way journeys. Kirch has written of the Lapita voyagers who "returned with the news of their discovery: there were indeed more islands to the east." In time, trade networks would develop, linking the islands of the Pacific in ways that today's commercial flight routes aspire to match.[11]

The Lapita paused for about a thousand years in these central island groups—Fiji, Sāmoa, and Tonga—where they developed into a new culture that anthropologists identify as Polynesian. The Polynesian culture, being descended from the Lapita, maintained the spirit of exploration inherited from its ancestors. After their long break from voyaging, some restless Polynesians loaded their canoes with breadfruit and the entire suite of cultivated plants that's come to be called "canoe crops" and set out to resurrect their voyaging past. They sailed east to find and settle Te Fenua ʻEnata (the Marquesas) and Rapa Nui (Easter Island), south to Aotearoa (New Zealand), and north to the Hawaiian Islands, all roughly a thousand years ago.

When Captain Cook explored the Pacific during the eighteenth century, more thoroughly than any of his European predecessors and most of those who would come later, he recognized that the Pacific Islanders shared cultural traditions and mutually intelligible languages. Cook witnessed examples of these common elements nearly everywhere he traveled between Australia and South America. When finally he ventured north of the Equator on what would be his last voyage and recognized an outpost of the same culture living and thriving in the islands we now call Hawaiʻi (Cook called them the Sandwich Islands, after John Montagu, Fourth Earl of Sandwich, one of his expedition's major sponsors), he entered into his captain's log what would become a famous question among Pacific historians:

"How shall we account for this Nation spreading itself so far over this Vast ocean?"[12]

Vast indeed. From the Bismarck Islands in the west to Rapa Nui in the east, the inhabited Pacific stretches more than 6,200 miles (10,000 km). The distance from Hawai'i in the north to New Zealand in the south is more than 4,300 miles (7,000 km) and, importantly for both native biodiversity and human-facilitated agriculture, this north-south reach takes in tropical and temperate climates on both sides of the Equator. Anthropologist Wade Davis has called the Pacific islands "the largest culture sphere ever brought into being by the human imagination."[13]

For many years, the answer to Cook's question was thought to be simple: Pacific Islanders spread themselves over "this Vast ocean" accidentally, parties engaged in nearshore fishing or short inter-island voyages repeatedly being blown off course and finding themselves haplessly and unintentionally washed ashore on some new island, until all the islands of Oceania were discovered and settled.

An alternative—and more believable—narrative is supported by the presence of *'ulu* and the rest of the canoe crops on nearly every island of the Pacific. These domesticated plants probably would not have been brought along on any voyage except those intended to settle a new island. The early twentieth-century ethnologists Craighill Handy and Elizabeth Green Handy made this point by asking their readers to reflect on the botanical evidence of breadfruit. If the voyages were not intended for colonization, they wrote, "breadfruit paste, cooked and packaged, would be stocked as provender, but not the sprouting root cuttings, wrapped with a ball of earth." Consider their point: When you take a vacation, you might bring along snacks for the trip, but do you bring your houseplants?[14]

The Hawaiian Islands were among the very last places on earth to be settled by humans. This beautiful archipelago was reached only as the early Pacific Islanders were in the final stage of expanding their domain. By the time the first twin-hulled voyaging canoe raised Haleakalā's massive red peak above the northern horizon, nearly every other island group in the Pacific had been found and settled. The techniques of navigating, voyaging, and settling new lands were well established and reliable. In his engaging account of the early human history of Hawai'i, *A Shark Going Inland Is My Chief*, Patrick Kirch proposed a way this last voyage of discovery may have played out.

In Kirch's tale, just over a thousand years ago, a young Marquesan *ariki*, or hereditary leader of his people, sets out to find new lands as a natural response to the inferior land grant his branch of the family tree had been allocated. This *ariki* was likely descended from a *younger* son of the first voyager to settle the Marquesas and, as such, did not have access to any of the more desirable valleys. Kirch laid out "two ways to overcome the lesser status conferred by birth," the first being to go to war against one's own relations in an attempt to "usurp the power of the more senior line." The second way to thrive despite an inferior caste was to "to discover and settle a new island where your own line would be paramount." Following a long tradition of younger brothers in the Pacific, this budding voyager began to prepare his canoe.[15]

The decision to sail north was novel, going against centuries of mainly eastward voyages of discovery. Perhaps the canoe's navigator had noticed flocks of *tōrea*, or golden plovers, flying northward and wondered where they would land. Maybe he knew he had no other direction to go. Pacific voyagers had already made landfall in South America. In doing so and doubling back, they had come to know the eastern and western limits of their vast sea. They had sailed at least as far south as New Zealand. In these eastern Pacific waters, however, no one had yet crossed the equator going north.

Since he couldn't know how long the voyage would be, the *ariki* would have called for heavy provisions. Not only would the crew need ample food for the journey, but if they were to settle on whatever new island they might find they would need the seeds and saplings with which to establish a new agricultural livelihood. Breadfruit root shoots and other plant cuttings would have been stowed, packed in equal parts Marquesan soil and earnest hope, that they might survive the journey to be planted in the earth of a new island.

Sweet potato and taro, dogs and pigs, men, women, and children, even rats and geckos all arrived in Hawaiʻi from the Marquesas. Not ʻulu. Kirch's narrative asks us to imagine a scenario in which "the delicate breadfruit saplings did not survive the salt spray, and withered before they could be established in the new land." I asked Kirch by email how he came to include this detail in his fictionalized yet scientifically informed account. He explained that, since archaeological evidence for the presence of ʻulu hasn't been found in the earliest postsettlement layers of the Hawaiian strata, a conservative hypothesis holds that the first settlers in Hawai'i did not arrive with breadfruit saplings—at least not any in a condition that favored cultivation. But the later evidence for *'ulu* is even more revealing as it supports the narrative of multiple back-and-forth voyages between Hawai'i and other Pacific islands: the *ariki*'s voyage was not one-way. It also lends credibility to the Hawaiian legend that *'ulu* was introduced not by the original Marquesan settlers, but later, from either Sāmoa or Tahiti.[16]

Here the field of linguistics provides corroborating evidence. In the Marquesas, breadfruit is called *mei*. If the Marquesan *ariki* who first settled the Hawaiian archipelago had introduced breadfruit, we might expect the Hawaiian and Marquesan words for the crop to resemble one another, like the greetings *aloha* (Hawaiian) and *ka'oha* (Marquesan) do. Instead, Hawaiians refer to breadfruit as *'ulu*, virtually the same word used in Sāmoa and a close cognate of the Tahitian *'uru*.

On each island they found, the Lapita, and later, the Polynesians, established societies, planted canoe crops including breadfruit, and created landscapes so that each island resembled the others. These settlers transformed their islands. They dug canals, constructed fishponds and taro patches, sculpted terraces, and meticulously selected and planted canoe crops, many of which, like breadfruit, were propagated vegetatively, rather than by seed, and were tended intensively by Pacific Islander farmers.

Consider the forethought that would have gone into that process as it played out across a vast inter-island stage. When a voyaging canoe arrived at a new island to be settled, each of those who had been charged with caring for the *'ulu* would carefully carry a tightly bound community—plant, soil, mulch, microbes, *mana*—to its new home, wrapped in leaves to keep moisture in, and would place this biological and geological community into a new hole, then water and chant.

Since breadfruit root shoots were transplanted with a small part of their soil intact, one could rightly say that the Hawaiian Islands are, in part, made from soil of Tahiti or Sāmoa, the locations far to the south from which the first *'ulu* came to Hawai'i. Tahiti and Sāmoa, in turn, are both made, to a small degree, of clods of soil from the western Pacific islands that still nurture *'ulu* groves today. As *'ulu* traveled, it diversified mainly in response to changing human preferences and growing conditions from island to island, giving us the amazing variety we see today across the Pacific. But the continuity made real by the movement of soil is reinforced by the inter-island connections evident in the shared breadfruit origin stories and myths found across this vast region. This same staple crop, grown in literally the same soil, owes its existence to an act of sacrifice in the stories handed down across generations, no matter where those stories are being told. I heard variations of the story of Kū giving up his life and becoming a breadfruit tree everywhere I traveled in the Pacific. The lesson of 'ulu, then, is a lesson of *kuleana*—a Hawaiian term referring to the reciprocity of rights and responsibilities—and an invitation to show gratitude to those who have taken on the responsibility to provide for others.[17]

Occasionally, islanders have forgotten their *kuleana* and have acted irresponsibly toward the gifts of the breadfruit tree. A story from Palau, in the western Pacific, illustrates this point and warns its listeners of the dire consequences they risk when they allow greed to eclipse gratitude. It tells of a woman—in some versions of the story her name is given as Milad, and she is sometimes identified as a demigoddess—from the island of Ngibtal, who taught the Palauans how to grow another of the region's staple crops: taro. Her reward for this act was a magical breadfruit tree that, when a branch was cut off, would allow the ocean to surge up through its hollow roots and out through the scar, bearing fish she could catch in a basket.[18]

The other residents of Ngibtal became jealous—why should Milad have a tree that provided both fruit and fish, with so little effort on her part? In generosity, she offered to share; in spite, they cut down the tree. Like Aesop's countryman who wasn't content to receive one golden goose egg each day, the Ngibtal islanders were disappointed to find there was no infinite store of fish inside the tree's trunk. Instead, the ocean rushed forth from the hollow stump and flooded Ngibtal, which remains underwater to this day. Pacific anthropologist Karen Nero has called this story "the most popular cultural image in Palau."[19]

On other Pacific islands, other breadfruit-related tales have arisen. For example, traditional Hawaiian mythology holds that 'ulu is one of the bodily manifestations—the *kino lau*—of Haumea, a goddess who, in human form, made frequent, rejuvenating trips to a hidden island, returning to bear children prolifically. The 'ulu tree, equally prolific, is associated with Haumea through the name, shared by both tree and goddess, Kāmeha'ikana, which means "a multitude of descendants." Like Haumea, 'ulu bears and bears again. And it is through Haumea that 'ulu has come to be associated with childbirth.[20]

A set of related traditions throughout the Pacific islands links birth with breadfruit. Chantal Spitz's novel *L'île des rêves écrasés* (The Island of

Shattered Dreams) is a multigenerational story of a Tahitian family engaging with an increasingly modern Pacific. Spitz tells the story of a new father, Maevarua, carrying his newborn son's placenta to a *marae*, an ancient site of traditional Tahitian spiritual life. Here, the new father prays to the "benevolent spirits of the marae," asking that their light and love be always upon his son, that they help him lead his son "along the path of his life's journey," and that they forgive him when, inevitably, he is weak.[21]

After praying to the traditional spirits, Maevarua then recites the Lord's Prayer—"Our Father, which art in Heaven . . ."—and, only then, "is doubly protected by the old order and the new." While his words may have spanned two orders, two religions, the next action taken by this new father was firmly connected to the order and the religion of his past. "With his child's placenta, he moves to the spot he chose with care . . . opens up the belly of the bountiful mother, places the placenta gently within, putting a young tumu 'uru on top of it, then replaces the soil."[22]

Spitz explained the logic behind the ritual of planting a *tumu 'uru*, a breadfruit sapling, above a child's buried placenta: the placenta had nourished the baby while in the womb, and "the tumu 'uru will nourish him through his life as a man." "This is the union of man with the earth into which he thrusts his roots, the union of the earth with man who makes his food spring forth from her belly," the novel continues, "For every birth, a bountiful tree is planted in the earth. With this gesture Maevarua carries the ancient soul of his people into the future."[23]

In some Pacific Islander communities it's the placenta; in others it's the umbilical cord. Anthropologist Solrun Williksen-Bakker explained how, in twentieth-century Fiji, in the face of so much modernization and cultural change, "the planting of the umbilical cord is one of the customs that young people today are still careful to perform." Of course, it is more difficult now that many babies are born in clinical settings with physicians and nurses moving in an out of the delivery room, bringing things in and taking things away. According to Williksen-Bakker, young Fijian mothers worry about the "risk of the umbilical cord of the child getting lost in the refuse containers with the dirty nappies. They beg the doctors and nurses to take care of it." Do these healthcare workers listen to the pleading mothers? "The Fijian staff do, non-Fijians sometimes forget."[24]

If the Fijian hospital staff have remembered their cultural responsibility to retain the umbilical cord, the new parents will carefully wrap it and take it home with their baby. The goal is to bring the cord to the parents' home island and to plant it there, along with a new breadfruit tree, which will grow to represent the child. Williksen-Bakker related the story of an old couple she knew, who lived in a village outside the modern Fijian capital of Suva. Looking out on their property, the couple "can contemplate 11 trees around their house. Each of these trees was planted together with the umbilical cord of one of their 11 children." Relatives can point to each tree and say to its child, "there you are." This burying of the umbilical cord under the young roots of a sapling, this melding of human and tree, symbolizes and makes real a relationship best described, in Williksen-Bakker's words, as "the treeness of life." The first act demonstrating life's treeness is the burial of the placenta or the umbilical cord with the planting of a tree soon after birth. The burial of the body after death is the last.[25]

While the traditions surrounding breadfruit's use in the Pacific are remarkably similar from island to island, the botanical diversity of the species itself is vast. Surveys of breadfruit diversity in the Pacific have found hundreds of varieties seemingly everywhere they've looked. Writing about the archipelagic nation of Vanuatu, the agricultural researcher Jean-Pierre Labouisse cited two surveys, one of which found sixty-five varieties of breadfruit growing in a single Vanuatuan village and, the other, sixty on an island less than 6.2 square miles (10 sq km) in area. Not sixty individual trees—sixty unique *varieties*. In the context of clonally propagated plants, a variety signifies a separate genetic line. Plant varieties can differ from one another slightly—like the various kinds of heirloom tomatoes—or dramatically like the wonderfully diverse and hard-to-believe-they're-all-related expressions of the *Brassica oleracea* vegetables: a single species with varieties representing such diverse vegetables as broccoli, brussels sprouts, cabbage, cauliflower, collard greens, and kale. Breadfruit varieties can be distinguished superficially by traits such as the color, size, shape, and taste of the fruits; the shape and size of the leaves; or the seasonality of the fruiting. To be sure, though, botanists examine the trees' genetics. Because nearly all the varieties of breadfruit arose due to selection by humans—or, in other words, through cultivation—each is referred to as a "cultivar": a portmanteau combining the words *cultivated* and *variety*.[26]

To witness the global breadth of breadfruit cultivar diversity, you could travel the world, combing the beaches of countless small atolls and bushwhacking up rocky gulches on hundreds of Pacific islands. Considering the possibility that new "creole" cultivars may have emerged in the

Caribbean over the last two centuries, your trip should probably take in at least a few of those islands as well. Along the way, you would meet members of a dispersed polyglot guild of farmers, orchard keepers, botanical gardeners, and amateur horticulturists. You could travel back in time, as it were, from Jamaica to Hawai'i to Sāmoa to New Guinea, witnessing the effects of artificial selection and natural adaptation on breadfruit as it journeyed eastward across the Pacific and then hopped to the Caribbean. Or you could stand on the shoulders of the botanical giants who have already made these journeys, returning with samples of breadfruit trees, leaves, and fruit to catalogue, describe, and cultivate all in one place: a library of breadfruit, so to speak.

If breadfruit had a Library of Alexandria it would be the Breadfruit Institute of the National Tropical Botanical Garden in Hawai'i and Diane Ragone would be its Zenodotus. For decades, until her retirement in 2022, Ragone, whom one journalist has called "a Jane Goodall of breadfruit," curated and directed the institute's main tree collection at Kahanu Garden on the island of Maui, where more than 150 distinct cultivars of breadfruit grow. This is the world's most diverse collection of living breadfruit trees. The first specimens, collected from around the Pacific and beyond, were planted in 1978 and Ragone's first major breadfruit research and conservation project resulted in the addition of more than a hundred new cultivars in 1991. To visit this living library, I rented a car at Maui's Kahului airport and took my family along on a drive down the famous "Road to Hana," pulling off at the dirt road that leads to the garden's entrance just before the small, picturesque town that serves as the route's namesake destination.[27]

When we arrived at the Kahanu Garden and Preserve, a woman at the visitor center radioed Mike Opgenorth, director of the garden and the day-to-day overseer of this priceless collection of breadfruit trees. As we waited in the rapidly alternating rain and sunshine for Mike to arrive, I asked whether many tourists come to see the *'ulu* grove. "Almost nobody," she told me. "They all come for the *heiau*." In addition to its breadfruit collection, Kahanu Garden is also home to the magnificent Pi'ilanihale Heiau, a ceremonial structure built in honor of Pi'ilani, the sixteenth-century chief who united all of Maui. When Mike arrived, he seemed genuinely pleased that someone had come with more interest in *'ulu* than the *heiau*. We set out for a walk through the grove.

Mike explained to me that the garden serves many roles: a site for both research and recreation, but also a repository of transplanted plants that may be endangered in their native environments. Whether due to development, sea-level rise, or the effects of war, some places in the Pacific with unique local breadfruit cultivars cannot support the trees in the numbers they once did. In these situations, Kahanu Garden serves in a role akin to the Svalbard Global Seed Vault, that heavily fortified Arctic storage site perched on a hill outside of the world's northernmost town, Longyearbyen, containing seeds from more than a million domesticated plant species. The website of the Crop Trust, one of the organizations responsible for the Seed Vault, calls it "the ultimate insurance policy for the world's food supply."[28]

Since breadfruit is normally propagated from air-layers, root shoots, and tissue cultures, its future is insured by continuous cultivation, not through the cold storage of dried seeds far above the Arctic Circle. While this may be more precarious, it maintains a closer relationship between people and trees than could be possible with a sterile seed vault. You could imagine an apocalyptic doomsday scenario in which survivors somehow make their way to Svalbard to procure the material to facilitate the renaissance of world agriculture, only to find that they no longer know what will sprout from the seeds they withdraw from the bank or how to care for plants that will grow. At Kahanu Garden, by contrast, you can see the very trees that are being conserved and, at the right time of year, taste their fruits.

Mike emphasized to me that the mission of Kahanu Garden does not overtly include food production. Still, the garden's trees are productive, and it would be irresponsible—against an *'ulu* grower's *kuleana*—simply to let their fruit fall and rot on the ground. Because of breadfruit's short postharvest shelf life, local distribution is the best way to avoid this wasteful fate. The garden has worked out arrangements to sell fruit to a few local Maui producers, including John Cadman, owner of the Maui Breadfruit Company in Kahului, and producer of Pono Pies. *Pono,* Mike explained to me, means "correct, righteous, or proper."

Leaving Kahanu Garden, my family and I drove back toward Kahului to John Cadman's small industrial kitchen, where Pono Pies are made. Walking in, I saw a surfboard hanging from the ceiling and heard Guns 'n' Roses blaring from a speaker. John met me at the door and, even before I sat down, placed an *'ulu*-based, mango-flavored, single-serving pie in my right hand and a small wooden spoon in my left. The label proclaimed

"Guilt free from Maui" and touted the locally sourced ingredients—breadfruit, macadamia nuts, mango, and honey—as well as the compostability of the packaging. John spoke about finding the appropriate scale for his business: big enough to be economically sustainable but small enough that he can still surf each morning before work. He told me about his efforts to source ingredients from both small backyard orchards and large growers like Kahanu Garden. The pie was delicious. I managed to get three more for my wife and children to try, but my mind kept returning to the immense variety of breadfruit growing down the road toward Hana.

Before contact with Europeans, the Hawaiian archipelago had one dominant breadfruit cultivar; it's simply called 'ulu or sometimes 'ulu maoli, meaning "authentic breadfruit." Botanists refer to this primary Hawaiian cultivar simply as 'Hawaiian.' These native Hawaiian 'ulu trees are still common today, but in recent decades a few tissue-cultured cultivars including 'Maʻafala,' 'Otea,' and 'Ulu fiti' have been widely distributed. In most cases, if you're eating 'ulu in Hawaiʻi, you can be reasonably sure that it's one of these few main cultivars. Except, that is, when you eat a Pono Pie. Since John sources so much of his 'ulu from Kahanu Garden, and since he'll take the fruit of any cultivar they'll sell him, when you eat a Pono Pie there's a chance you might be eating breadfruit of an extremely rare cultivar—rare not just in Hawaiʻi but, sometimes, in the world.

Some people are entranced by uncommon and "exotic" foods. As a student at Cambridge, Charles Darwin was a founding member of the Glutton Club, which met regularly to dine on "birds and beasts which were before unknown to the human palate." Later, during his famous and foundational voyage aboard the *Beagle*—recall that Darwin's official role was not originally that of a naturalist but as the "messmate" or dining companion to the captain, Robert FitzRoy—he took the opportunity to expand his palate further, consuming such Georgian-era novelties as bananas, tamarinds, and "a profusion of oranges." After his round-the-world voyage Darwin would remain in England for the rest of his life, occasionally satisfying his "taste for exotic fruit" by sampling some of the tropical produce that his friend, Joseph Dalton Hooker, would send from the greenhouses at Kew Gardens.[29]

With Pono Pies, you can satisfy the desire to eat something rare—a kind of breadfruit that may exist only at Kahanu Garden and, perhaps, on one other Pacific island—without endangering a species or yourself. In other words, you can expand your palate with unusual and remarkable foods in a way that actually is *pono*.

CHAPTER IV

Awful and Lovely

In May 1513, the Portuguese explorer Jorge Álvares arrived at China's Pearl River estuary, landing on an island he called Tamão. The present-day Chinese island corresponding to Tamão is unknown, but it has been suggested that while exploring the area, Álvares may have become the first European to view the Pacific Ocean. About four months later, the Spanish conquistador Vasco Núñez de Balboa reached the top of a hill in the Darién region of Panama and took in his first view of the eastern edge of the same great ocean. Between these two wandering Iberians, Álvares and Balboa, on all the islands of the Pacific, life went on without anyone taking notice. The great changes would only come when Europeans ventured out to claim and colonize these islands and their peoples.[1]

While the image of Álvares and Balboa standing on the Pacific's western and eastern shores almost simultaneously in 1513 can be appealing to scholars of exploration, historian Charles Nowell has compiled ample evidence to suggest that their so-called codiscovery was both insignificant and possibly preempted. On its insignificance, Nowell reminded his readers of the millions of Native peoples living on the coasts and islands of the Pacific who already knew the ocean well. Even on the narrow topic of *European* knowledge of the Pacific, Nowell referenced multiple explorers, dating back to Marco Polo, who may have preceded Álvares. It was through the reports of Álvares and Balboa, however, that Europeans began to think of attempting their first Pacific crossing.[2]

Fernão de Magalhães, better known in English as Ferdinand Magellan, was the first European to cross the Pacific Ocean, and the first non–Pacific Islander, therefore, to understand its expanse. Magellan's fleet left Seville in August 1519. After crossing the Atlantic and rounding the southern tip of South America through the straits that would later bear his name, Magellan took more than three months to cross the Pacific, having unwittingly chosen a route that would bypass nearly all the great civilizations of Oceania. Instead, the ships passed by only two small, uninhabited islands, neither of which offered a safe anchorage, so both were simply noted and passed by. Magellan's first landfall since leaving South America was Guam. The voyage was plagued by violence, both among the crew and directed toward the islanders they would encounter in the Marianas, the Moluccas, and finally the Philippines, where Magellan himself was killed. The small remnant of sailors who survived to return to Seville in Magellan's absence resembled "the face of every sailor's nightmare," owing to their severe malnutrition, scurvy, and the rest of the horrors they had experienced.[3]

Despite the suffering and loss encountered during this first European Pacific crossing, many later voyages would follow in Magellan's wake. Europeans went to the Pacific to explore and to escape, to colonize and to Christianize. Throughout this age of exploration, each Pacific island had its own first contact with Europeans. Some of those encounters were peaceful; many were not. All resulted in major changes to the cultures and ecosystems of the Pacific islands. In these encounters, it is illuminating to see how the Europeans imagined the unfamiliar environments and bountiful produce of these islands, including, especially, breadfruit: sometimes as a savior, making harsh island life possible; sometimes a curse, killing industriousness and fostering sloth among Pacific Islander communities as its abundant productivity retarded the development of a respectable work ethic; and almost always as a curious and "exotic" symbol of the near-mythical "South Seas" and the wholly unfamiliar way of life that seemed to exist there.

Before going any further with this consideration of European–Pacific Islander contact, I need to acknowledge the undue weight that the first instances of such interactions have been given in histories of non-European cultures. Too often, written histories begin with the European "discovery" of a place and its people; all that happened before is either summarized in a preface or merely left out. For example, the opening line of Gavan Dawes's classic history of Hawai'i, *Shoal of Time*, sets the book's beginning "at dawn on January 18, 1778" with Cook's arrival. Woodrow Wilson, who began a career as an academic historian before entering politics and eventually becoming the twenty-eighth president of the United States, dedicated only the first thirty-three pages of his five-volume *A History of the American People* to a chapter titled "Before the English Came." Even then, the chapter all but ignores Native American history as it deals almost exclusively with the French and Spanish exploration of North America as a prelude to the settlement of the continent by those whom Wilson called "proper men and proper means."[4]

We know, though, that histories do not begin when Europeans arrive on the scene; most histories have no clear beginning point at all. Threads just disappear into the murk of time to be teased out and hinted at by archaeologists and mythmakers. The future may indeed lie before you, but we encounter time like the passenger "riding backwards on a train" in James Hoch's poem by that title. The future approaches from out of sight; we may have an idea of where we're supposed to be going but our progress only emerges as we pass the stations, villages, and fields along the way. Our view takes in the past, but its furthest reaches disappear as we round each bend or emerge from each tunnel.[5]

The history of breadfruit in the Pacific was already long and nuanced—receding into the past, disappearing far beyond the view of any living person's backward-facing memory—when first it gained a written description in a European language. That language, it would turn out, was a Portuguese-accented Spanish. The description was produced during the voyage of Álvaro de Mendaña y Neira in the service of the Spanish Crown, written, or dictated, by Pedro Fernandes de Queirós—the Évora-born navigator who would take command of the *San Gerónimo*, Mendaña's ship, after the captain's death in the Solomon Islands. Breadfruit had been part of a complex Pacific ecosystem for centuries before Europeans learned about it, but its next journey—its introduction to the Caribbean—would depend heavily upon the way Europeans imagined it.

In 1595, the Mendaña expedition sighted the mountainous island of Fatu Hiva in the eastern Pacific and anchored off the nearby island of Tahuata, taking it upon themselves to rename the entire archipelago Las Marquesas de Mendoza, after the Viceroy of Peru. It was in the Marquesas that Queirós produced the first known written account of breadfruit. There, he described trees that "yield a fruit which reaches to the size of a boy's head. Its colour, when it is ripe, is a clear green.... The rind has crossed scales like a pineapple, its shape not quite round, being rather more narrow at the end than near the stem.... It has no core nor pips, nor anything uneatable except the skin, and that is thin. All the rest is a mass of pulp when ripe, not so much when green. They feed much upon it in all sorts of ways, and it is so wholesome that they call it white food." After recording that description, Queirós and the rest of Mendaña's crew murdered two hundred Marquesans and continued sailing westward across the Pacific.[6]

Mendaña was an explorer whose embrace of violence approached the level of piracy. Another important figure in the history of European impressions of the Pacific, and of breadfruit, was William Dampier—an English pirate whose curiosity, observational skills, and prolific writing approached the level of an explorer. Dampier was a sailor aboard the *Cygnet*, captained, poetically, by Charles Swan, a reluctant privateer who led raids against Spanish holdings along the Pacific coast of the Americas, from Peru northward to Mexico. In 1686, Swan decided to follow the Spanish galleons on their regular transpacific Acapulco-to-Manila runs, hoping to overtake a ship and claim its cargo. The *Cygnet* succeeded in making the Pacific crossing but failed to capture any Spanish vessels.

Instead, Swan's crew called at Guam where Dampier first encountered a fruit that, in his words, "the Natives of this island use . . . for Bread." Because there is no earlier written example of *bread-fruit* than Dampier's 1697 account, and because he qualified his use of the term—"The Breadfruit (as we call it)"—scholars credit Dampier with giving us the English word. *Breadfruit* certainly has a more evocative ring to it than Queirós's "white food" and Dampier, whom one biographer called "a humane man in a not very humane age," seems more worthy of being remembered for his botanical and linguistic contributions than the murderous Mendaña.[7]

About fifty years after Dampier's unsuccessful attempt at piracy in the Pacific, George Anson of the Royal Navy set out on a round-the-world voyage with the same goal: to raid the Spanish possessions and, if possible, to capture a Manila galleon. On this latter goal, unlike Swan and Dampier, Anson was successful, having overtaken and claimed *Nuestra Señora de Covadonga*, referred to among pirates and privateers as "the Prize of all the Oceans." Anson also encountered breadfruit, in his case, for the first time on the island of Tinian in the Marianas. "A fruit peculiar to these Islands," he called it, "which served instead of bread." Anson went on to describe how popular breadfruit was among his crew: "for it was constantly eaten by us during our stay upon the Island instead of bread, and so universally preferred to it, that no ship's bread was expended." Anson then provided a botanical description of the breadfruit tree and of some of the culinary uses of the fruit. He cited Dampier, as well as the botanist John Ray, as having previously described breadfruit. Ray, in fact did not describe it in his three-volume *Historia Plantarum*, but an appendix was added to a later printing, written by the botanist Georg Josef Kamel, which included a description of breadfruit along with other plants of the Philippines.[8]

Mendaña, Queirós, Dampier, and Anson all added to Europe's knowledge of the Pacific islands. But the person who made what may have been the most lasting contribution to the European idea of the region, and of breadfruit's role there, was the explorer Louis-Antoine de Bougainville. As commander of the first global circumnavigation by a French fleet, Bougainville visited Tahiti in 1767. It's fair to say that he liked what he saw. Bougainville's account of the voyage described the crew's amazement at the people, landscapes, and produce of the island, not always with the purest intentions. Among other lecherous remarks, Bougainville recorded in his journal an incident in which, "we also saw four or five women, of whom two were fairly passable but the others abominably ugly. Three had a breadfruit leaf to cover themselves and the others naked as in Nature; unfortunately it was the ugly ones."[9]

Both the journal Bougainville kept during his voyage and the book he published afterward offered very little new information about breadfruit, beyond what Queirós, Dampier, and Anson had already written, except, I suppose, its utility in clothing the bodies of "fairly passable" women. Instead, he provided Europeans with an idea that would survive to the present day. Bougainville created the concept of the Pacific islands as Paradise.[10]

He did so by linking Tahiti to the classical age, in particular, the Greek myth of Aphrodite, goddess of love, beauty, and fertility—Venus to the Romans and also known as Kytherea, "since from Kythera she was come." The Greek island of Kythera, more commonly rendered in English as "Cythera," was closely identified with Aphrodite. It was this idyllic island that Bougainville had in mind when he renamed Tahiti "la Nouvelle Cythère," or the New Cythera.[11]

Some translators of Hesiod have rendered the poet's description of Aphrodite, "aidoié kalé theós" as "an awful and lovely goddess." The same uneasy combination of *awful* and *lovely* has been applied to the islands of the Pacific in general, and to Tahiti in particular, since Europeans first began exploring these antipodal places. Early European contact with Pacific

Islanders was marked by a mix of envy and disdain for what they viewed through biblical lenses as perhaps a remnant of pre-Fall humanity. The landscapes and lagoons of the Pacific islands were lovely to look at and the productivity of the fields and groves seemed to show that the islanders were unencumbered by the divine affliction with which Adam and Eve were banished from their garden to a place of cursed ground where God punished them saying, "in the sweat of thy face shalt thou eat bread."[12]

To Europeans like Bougainville, it appeared that the Pacific Islanders did *not* have to eat bread in the sweat of their faces, nor was their ground cursed, for in it grew the breadfruit among so many other bountiful plants. In Tahiti, Bougainville wrote, "*Je me croyois transporté dans le Jardin d'Eden*" (I believed myself transported to the Garden of Eden). After Bougainville, European descriptions of the Pacific islands consistently referred to breadfruit's supposed facilitation of an easy, labor-free, paradisical life, as these lines from Byron exemplify:

> The bread-tree, which, without the ploughshare, yields
> The unreaped harvest of unfurrowed fields,
> And bakes its unadulterated loaves
> Without a furnace in unpurchased groves,
> And flings off famine from its fertile breast,
> A priceless market for the gathering guest.[13]

Like Bougainville and Byron, Captain Cook mused that the inhabitants of Tahiti may have been "exempted from the curse of our fore fathers," observing that they did not have to "earn their bread by the sweat of their brows," but rather, that "benevolent nature hath not only provided them with necessaries but many of the luxuries of life; Loaves of Bread, or at least what serves as a most excellent substitute grows here in a manner spontaneously upon trees." Of course, Cook and the rest of these early European explorers failed to acknowledge the generations of agricultural work that had preceded their arrival in the region: digging canals, fishponds, and taro patches, sculpting terraces, and meticulously selecting and planting canoe crops, many of which, like breadfruit, were propagated vegetatively, rather than by seed, and tended intensively by Pacific Islander farmers.[14]

As Cook explored the Pacific, he traversed temperate and tropical waters between the Arctic and the Antarctic. He encountered high, volcanic islands and low, coral atolls. Cook and the naturalists with whom he sailed observed the remarkable adaptability of both people and plants in these diverse environments. Breadfruit trees seem to prefer to be planted in the deep, fertile, well-drained soils of the equatorial lowlands with plenty of rainfall and protection from wind and saltwater. But even today, horticulturists marvel at breadfruit's tolerance of such environmental challenges as high elevation, relatively cool temperature, acidic or saline soil, frequent strong wind, and variable precipitation.[15]

But humans are even more adaptable. When voyaging Polynesians first reached the two large islands they would call Te Ika a Māui and Te Waka a Māui ("Māui's fish" and "Māui's canoe") about eight hundred years ago, they would have recognized an environment different from anything they or their ancestors had seen before. People arriving in New Zealand today with knowledge of other Pacific islands continue to draw the same contrast. While most islands settled by the Lapita and their descendants are fully tropical, New Zealand is both temperate in latitude and high in elevation. The New Zealand climate, which supports both the giant subtropical kauri trees in the north and the glaciers of the Southern Alps, is like nothing anywhere else in the insular Pacific.[16]

One nineteenth-century English immigrant to New Zealand sought to instruct his fellow colonists on the matter of climatic variability among the Pacific islands. Charles Hursthouse, in a book that refers to New Zealand as "the Britain of the South," declared that "in making choice of any new land, 'good climate' should unquestionably rank as our first requirement." He went on to clarify that, "by good climate, I mean a bracing, temperate climate," and asserted, quite Britishly, that "a climate can be too fine." Then, perhaps thinking back to something he may have read in an explorer's journal, Hursthouse concluded that, "where there is perpetual sunshine and serenity, man degenerates into an emasculated idler. . . . In Tahiti and the Paphian coral isles of the South, the soft islanders are scarce equal to greater labours than plucking fruit and eating it." With this, Hursthouse put to words Europeans' contradictory views on breadfruit. Its perceived ease of cultivation and abundant productivity contributed to the islands' resemblance to Paradise, but these very same attributes made the inhabitants weak and lazy. Thus, steeped in what would come to be called the "Protestant work ethic," it was the duty of any self-respecting European to disdain the breadfruit for its complicity in the degeneration of mankind. This idea was captured perfectly by Edna Dean Proctor's poem "The Glory of Toil," which declared,

> Better the lot of the sunless mine, the fisher's perilous sea,
> Than the slothful ease of him who sleeps in the shade of his
>   breadfruit tree;
> For sloth is death and stress is life in all God's realms that are,
> And the joy of the limitless heavens is the whirl of star with star![17]

This idea—that the inhabitants of the tropical Pacific islands were languid, averse to labor, and disinclined to expend any unnecessary energy, along with the logical inverse that people from more temperate regions of the world were hard-working and industrious—became pervasive as colonialism expanded throughout the tropics and Europeans began to demand work from unwilling local people. Sociologist Syed Hussein Alatas has shown in *The Myth of the Lazy Native* how these stereotypes and misconceptions served to justify the taking of colonial possessions, owing to the belief that the supposed laziness of the Native peoples corresponded with an inability to govern themselves. Just as Aphrodite was described as a goddess both awful and lovely, Tahiti, the Pacific island most closely associated with her, and by extension the entire region, fell under the same tenuous descriptors. Europeans found the Pacific islands to be lovely for the physical beauty of their inhabitants and the productive bounty of their land, but awful for the assumed listlessness and lack of industry that these generous environments enabled among these beautiful people.[18]

Even Alexander von Humboldt, the progressive (for his time) eighteenth- and nineteenth-century geographer, fell under the spell of Alatas's "myth" as he described tropical countries where "in the midst of abundance, beneath the shade of the plantain and breadfruit tree, the intellectual faculties unfold themselves less rapidly than under a rigorous sky, in the region of corn, where our race is engaged in a perpetual struggle with the elements." Humboldt blamed the rich soil and "vigour of organic life" that he found in the tropics for "retard[ing] the progress of nations toward civilization."[19]

When Hursthouse used the term *Paphian* in reference to the Tahitians, he directly recalled the myth that gave us Bougainville's Nouvelle Cythère, for Paphos is a city on the southwest coast of Cyprus, where Aphrodite was said to have first set foot before moving on to Cythera to become associated with that island as well. At Paphos once stood "the most famous sanctuary of the goddess," whose identification with the city was so strong that she came also to be called—already not short on alternative appellations—"the Paphian," or "Paphia."[20]

In contrast to the "too fine" climate of the South Pacific and the apparent shortcomings of the cultures that developed there, Hursthouse praised New Zealand, which, he said, possessed "a climate most congenial to all pastoral and agricultural pursuits." What's more, in his environmentally deterministic way of thinking, Hursthouse considered the prototypical example of New Zealand's Māori people to be "no . . . soft Paphian islander nourished on bread-fruit pap; but a black Scandinavian of the south, a very clever and ferocious savage, who ate the dog and the shark, who drank blood, who scoffed at death, and who any day in the year would fight his enemy to the death for the prize of the corpse to feast on."[21]

The Māori are certainly neither Black nor Scandinavian, but they did in fact eat dog, shark, and even, occasionally, human flesh. More to the geographical point underlying Hursthouse's description, the Māori adapted to an environment that was very different from the tropical islands their ancestors knew. Lying far south of the Tropic of Capricorn, New Zealand is in a temperate climate, unlike the tropical islands from which the ancestors of its Native population came. Wellington is the southernmost national capital in the world. Invercargill, on the south shore of New Zealand's South Island, sits at 46° south latitude, farther south than Montréal is north. Even Cape Reinga, near New Zealand's northernmost point, sits as far from the equator as Los Angeles. And, as is true for northern latitudes from Los Angeles to Montréal, breadfruit does not grow in New Zealand.[22]

Given the widespread view that the tropical climate of a place would have degenerative effects on the character and intellect of the people who lived there, early Europeans generally thought of themselves as visitors to the tropical Pacific, not as permanent settlers. Those few who found themselves existing as castaways and decided to integrate into Pacific Islander societies were occasionally "rescued" against their will by the crews of other European ships who could not believe, or understand, that anyone would choose to stay in these islands so very different from their homes.[23]

Among the first exceptions to this rule, as might be expected, were the artists. Notable Europeans like Paul Gauguin and Robert Louis Stevenson who moved more or less permanently to the Pacific islands had to contest the assumption that they had sought mere escapism or worse, that the supposedly degenerative effects of island living, the labor-free subsistence offered by breadfruit and the rest of the islands' bounty, had affected them as it was thought to have shaped the Indigenous cultures. Stevenson's grave in Sāmoa stands in a clearing bordered by breadfruit trees atop Mount Vaea. Arriving after a steep hike to the summit, I found a large concrete tomb with small offerings of folded leaves and candles placed around its perimeter. On one side I read the inscription, seemingly seeking to justify Stevenson's presence there: "Here he lies where he longed to be; / Home is the sailor, home from the sea." Similarly, Gauguin would be remembered as an artist who sought to escape: first from the standards of European artistic style and then, taking up residence in French Polynesia, as one who sought to escape from European society itself.

But Stevenson, Gauguin, and a host of other European and American artists who spent more than a brief holiday in the Pacific produced some of the most original works of art and literature of their times. Their work used breadfruit and other produce of the Pacific islands to conjure a sense of Paradise not unlike Bougainville's idea of Nouvelle Cythère. Fallen breadfruit leaves are scattered in the foreground of Gauguin's *Te raau rahi (The Big Tree)* and living leaves stand out from the foliage in the background of *Ia Orana Maria (Hail Mary)*. Stevenson wrote of "breadfruit gilt in the fire" and of the "deepest pit of popoi" in his *Ballads*.[24]

Considering other authors, the illustrated 1921 version of *Peter Pan* included details about the Lost Boys' diet, stating specifically that "their chief food was roasted bread-fruit, yams, coconuts, baked pig, mammee-apples, tappa rolls and bananas, washed down with calabashes of poe-poe." The beverage with which the Lost Boys washed down their meal is likely a reference to the Marquesan delicacy made of pounded, fermented breadfruit called *popoi*, which I had observed being made on Nuku Hiva. *Peter Pan*'s author, J. M. Barrie, had apparently misunderstood the consistency of this dish, which is a thick dough or paste, instead considering it to be a liquid. Literary scholar David Park Williams has called this menu "straight out of *Typee*."[25]

In *Typee: A Peep at Polynesian Life*, Herman Melville's first book, the author described *popoi* ("poee-poee" in his phonetic spelling) as a "staple article of food among the Marquese islanders," and one that is "manufactured from the produce of the bread-fruit tree," likening "its plastic nature [to] our bookbinder's paste," and describing it as being "of a yellow colour, and somewhat tart to the taste." The meal in which Melville's narrator, Tommo, first encountered *popoi* also included coconut water, served in the "natural goblets" of coconut shells, and Melville described how refreshing this drink was in the tropical warmth. It is likely, if Barrie was indeed inspired by *Typee*, that he confused the liquid coconut water and the pasty *popoi*, casting the latter in the role of a beverage in his Lost Boys' meal.[26]

How did Melville describe *popoi* so accurately? He knew it personally, having spent the summer of 1842 on Nuku Hiva, living mainly in Taipivai, the same valley where Hervé Ah-Scha showed me how to make *popoi*. The title of his book, *Typee*, is based on a shortened form of the name of the valley, Taipi.

At the age of twenty-one, Melville signed onboard the whaling ship *Acushnet* as a "green hand," or a crewmember lacking experience. For eighteen months he participated in whaling while he observed the people and places that would fill his novels to come. Melville would later write in *Moby-Dick*, undoubtedly speaking autobiographically through his character, Ishmael, that "a whale-ship was my Yale College and my Harvard." After eighteen months of whaling, the *Acushnet* entered the port of Taiohe in the Marquesas. Here Melville admitted, through Tommo, that he had decided to desert his post, to drop out of school, as it were. "I chose rather to risk my fortunes among the savages of the island than to endure another voyage," Tommo confessed.[27]

Melville mentioned breadfruit only once in *Moby-Dick*: as a side dish to a cannibalistic meal described by the harpooner, Queequeg, who hailed from the fictional South Pacific island of Kokovoko. Queequeg described the event as "a great feast given by his father the king . . . wherein fifty of the enemy had been killed by about two o'clock in the afternoon, and all cooked and eaten that very evening." The corpses were "garnished round like a pilau, with breadfruit and cocoanuts; and with some parsley in their mouths."[28]

In *Typee*, however, breadfruit is a regular—and far more palatable—part of Melville's narrative. In the 1920 edition of the book, one chapter is even titled "Bread-fruit." Here Melville provided a detailed description of the tree itself—"a grand and towering object, forming the same feature in a Marquesan landscape that the patriarchal elm does in New England scenery"—and then proceeded to discuss several ways the fruit can be prepared. The first, the simplest, is the classic roast breadfruit, blackened atop a fire and eaten "in its purest and most delicious state." Melville also described *ka'aku*, the Marquesan dish made from pounded breadfruit paste mixed with coconut milk, which he called "a dish fit for a king."[29]

AWFUL AND LOVELY

Nearly two hundred years after Melville's time in the Marquesas, on Hiva Oa, another of the islands in the archipelago, I was served *ka'aku* at dinner in the guesthouse where I stayed. Its texture was like panna cotta, its color like crème brûlée, and it was surrounded on the plate by puddles of excess, unabsorbed coconut milk. The taste was sweet but dominated by the smoke of the wood fire upon which the breadfruit had been roasted. Underneath the smoke I could clearly taste the starchy breadfruit and sweet coconut. As I ate, I was unsure whether I should consider the *ka'aku* to be my dessert but then the chef brought out an enormous platter of crêpes with whipped cream and jam, so I guessed the *ka'aku* had simply been intended as a sweet and hearty side dish.

Breadfruit also makes an appearance in *Twenty Thousand Leagues Under the Sea*. When Jules Verne's *Nautilus* surfaced for a few days on land, the crew sought out fresh food. They were rewarded, the narrator, Pierre Aronnax, explained, by "one of the most useful products of the tropical zones. . . . the bread-fruit tree." One crewmember, Ned Land, pleaded with the captain: "Master," he said, "I shall die if I do not taste a little of this bread-fruit pie." Permission was granted, and Land set about preparing roast breadfruit. Verne accurately captured the cooking process, describing how Land took breadfruits and "placed them on a coal fire, after having cut them in thick slices, and repeating: 'You will see, master, how good this bread is. . . . It is not even bread,' added he, 'but a delicate pastry.'" "After some minutes," the story continues, "the part of the fruits that was exposed to the fire was completely roasted. The interior looked like a white pastry, a sort of soft crumb, the flavor of which was like that of an artichoke." And here Aronnax declared his own opinion, "it must be confessed this bread was excellent, and I ate of it with great relish."[30]

In art, literature, and the journals of some early European explorers, breadfruit and the islands on which it grew shared Aphrodite's lovely attributes. But in other contexts, the supposedly paradisical nature of life in these tropical islands of the Pacific, nurtured as it was by breadfruit, which allegedly grew spontaneously upon trees without human intervention, conjured the goddess's other side, her awful nature. And this inspired not desire or envy on the part of the Europeans but abhorrence.

What Anson, Bougainville, Cook, and Dampier observed, what Humboldt and Hursthouse detested, what Gauguin painted, and what Barrie, Melville, and Verne described, others attempted to destroy. According to literary scholar Peter Kitson, a scheme was devised, the origin of which

Kitson traced to the poet Samuel Taylor Coleridge, "to mend the Tahitians by extirpating the bread-fruit tree from their island and making them live by the sweat of their brows." This model of subjugation by breadfruit eradication was not original to European colonizers. According to James Morrison, who served in the Pacific aboard HMS *Bounty* during the late eighteenth century, the precolonial elite of Tahiti forbade the residents of the nearby atoll Teti'aroa to grow breadfruit, specifically so that the Tetiaroan people would have to rely upon the Tahitians for subsistence. Morrison wrote, referencing Teti'aroa, that Pōmare, then head of Tahiti's ruling family, "keeps the Inhabitants in subjection by keeping them from Planting the Breadfruit or other Trees and suffers nothing to grow there except a few Tarro for His own use."[31]

Teti'aroa, once the private island retreat of actor Marlon Brando, is now home to The Brando, a luxury resort. Morrison's account of the atoll's breadfruit history was confirmed to me as recently as April 2022, in an email exchange with Teva Beguet, a ranger at the Tetiaroa Society—an organization designated "as the environmental steward of Teti'aroa"—who wrote, "According to the history of the atoll, it used to have breadfruit trees . . . but these trees were cut down during Pōmare's reign. Now the only breadfruit trees on the atoll were planted by the resort gardeners."[32]

This seemed worth investigation, so, a few months after corresponding with Teva, I boarded a small plane at Tahiti's Fa'a'ā International Airport for the short hop over to Teti'aroa. I traveled as a guest of the Tetiaroa Society, not The Brando, and I was soon to learn about the near-absolute segregation between these two entities. The "TS," as it is called on the island, occupies the northern portion of Onetahi, one of the atoll's motus, or islets, and is centered around a tidy but spartan laboratory/bunkhouse compound. The Brando, by contrast, sprawls across the southern half of the same motu and is regarded as one of the world's most exclusive resorts. When Barack Obama's second term as U.S. president ended in 2017, he retreated to The Brando for nearly a month to write his memoir.[33]

After landing on Teti'aroa, I, as a visiting researcher, was quickly whisked away from the resort guests' arrivals line. They were welcomed with leis and ukulele, I joined The Brando's staff members under a thatched roof where Antonin Fioretti, the TS Ecostation Manager, met me in an electric golf cart. My tour was to begin immediately.

Our first stop was the organic garden where fruits and vegetables are grown for the resort's Michelin-starred chefs to use in their award-winning cuisine. Two young, productive, breadfruit trees grew in this garden, but the gardener, Gilbert, emphasized that he had planted these only recently at the resort chefs' request; they had not escaped Pōmare's eradication. I borrowed one of the TS's many beach-cruiser bicycles—the tires intentionally kept flat to accommodate riding on loose sand—and spent the rest of the day biking around Onetahi, searching the forests for breadfruit trees gone feral or otherwise unnoticed. Those in the garden were all I found. When I sat down later to talk with Tihoni Maire, head guide for the resort, I learned an interesting bit of information about Pōmare's motivation for cutting down Teti'aroa's breadfruit trees. Morrison—the *Bounty* sailor—and Teva Beguet of the Tetiaroa Society had both been right that the Tahitian ruler removed the trees to keep "the Inhabitants in subjection," but, according to Tihoni, this subjection went beyond mere nutritional dependence upon Tahiti. Recalling the association between breadfruit trees and childbirth, best illustrated by the burial of a newborn's placenta or umbilical cord and the planting of a breadfruit tree on top, Tihoni explained that, when Pōmare cut down the breadfruit trees of Teti'aroa, he also severed the umbilical connection between the island and those people who had been born there. Without a tree to mark the place where your birth had been consecrated, you become rootless; you drift.

But one thing about breadfruit trees is that they are resilient. Whether chopped down by a power-hungry ruler or blown down by a hurricane, they tend to grow back. If a tree is cut at ground level, its roots left intact, new root shoots will often arise from the ground. To a person whose own birth had been marked by a breadfruit tree, it must offer some degree of solace to see your tree grow back after a disaster. It must remind you of your own resilience. Even though the people whose births had been represented by the breadfruit of Teti'aroa back during Pōmare's reign had long since died, I wanted to find the trees' offspring, their *keiki* as the Hawaiians would say, to affirm the strength of the human spirit.

The next day, after a night spent in the bunkhouse, I took a boat with a TS ranger to another of the atoll's motus, Rimatu'u, to see the ruins of an old village where workers had lived and produced copra—dried coconut flesh from which coconut oil can be pressed—during the early twentieth century. It was on Rimatu'u that I hoped to find pre-resort breadfruit trees, since the botanists Marie-Hélène Sachet and Ray Fosberg had included the species in their 1983 survey of the atoll, conducted at the behest of Marlon Brando himself. I spent the day tramping along the overgrown trails of Rimatu'u, looking for the distinctive large, lobed leaves of the breadfruit tree. There were plenty of coconut palms and pandanus. Ferns sheltered skittering land crabs. The homes and other buildings of this former village stood collapsing, being absorbed back into the tropical forest. I found no breadfruit.[34]

While breadfruit was "only seen at Rimatuu village" in 1983, Sachet and Fosberg referenced a botanical survey published a decade earlier that had recorded its presence "around camp-sites on Onetahi and Tiaraunu islets." I had already crisscrossed Onetahi by bicycle but had not yet seen Tiaraunu, so I borrowed a TS kayak, loaded my gear and food for the day into a lidded, five-gallon bucket labeled "Rodent Bait," and paddled across the channel to continue my search. The campsites were easy to find: little coves in the margins between forest and beach, some with tents pitched for researchers' overnight stays. I beached the kayak and walked from one campsite to the next, scanning the forest edge for breadfruit trees. Juvenile blacktip reef sharks kept pace with me, swimming just offshore. I walked and kayaked the perimeter of the Tiaraunu motu. Still no breadfruit.

That evening, before leaving Teti'aroa the next day, I borrowed another bicycle and rode across the airstrip to a section of Onetahi called, inexplicably, "Mexico." Here, across the runway, a small village of shipping-container cottages had popped up alongside the few remaining hexagonal, wooden bungalows that Brando had constructed for his guests while the atoll was still in its pre-resort, private-island days. The shipping container cottages house some of the resort staff. One of the bungalows is the on-island home of Frank Murphy, executive director of the Tetiaroa Society. Another belongs to Teihotu Brando, Marlon's son.

Marlon Brando had first seen Teti'aroa while filming the 1962 film *Mutiny on the Bounty* in Tahiti. In the movie, Brando played Fletcher Christian, the sailor who led the mutiny and oversaw the process of throwing the ship's cargo of more than a thousand potted breadfruit saplings overboard. Of Brando's dozens of feature films, *Mutiny on the Bounty* may have had the greatest impact on his life. During filming, he wrote in his memoir, "every day as soon as the director said, 'Cut' for the last time, I ripped off my British naval officer's uniform and dove off the ship into the bay to swim with the Tahitian extras working on the movie." Brando went on to write of the Tahitians he met, "I grew to love them for their love of life."[35]

Tarita Teri'ipaia, the actor who played Mauatua, Fletcher Christian's love interest, became Brando's third wife and Teihotu's mother. In 1966,

Brando purchased Teti'aroa and built several structures that would serve as a vacation home for his family. After Marlon Brando's death in 2004, and before the development of The Brando resort, Teihotu lived on Teti'aroa alone with his wife and children, surviving, according to a journalist for *Maxim* magazine, "on the fish he spears, the fruit he picks, and whatever provisions his occasional visitors can bring from Tahiti." Now he lives on the island part-time and occasionally drives boats for the Tetiaroa Society and resort guests alike.[36]

I found Teihotu sitting on his lanai, enjoying a Heineken with a friend, and I stopped my bike to say hello. When I told him about my interest in breadfruit he stood, excited. "Breadfruit? I just planted one!" he said, gesturing toward a semi-shaded patch of grass between the airport runway and his bungalow. Teihotu set down his beer and walked over to show me, stabilized by a pile of coconut husks at its base, a diminutive breadfruit tree, growing no higher than our knees. But its size didn't matter. The tropical sun and island soil would feed this sapling, and it would soon grow quickly into a tree to provide shade for Teihotu's bungalow and fruit for his table. More than two hundred years after Pōmare chopped them all down, another breadfruit tree had taken root in Teti'aroa's soil, this one planted by a Brando.

CHAPTER V

The Wondrous Food of the Land

Traditional Pacific navigation—the body of skills and knowledge that allowed Teti'aroa, Tahiti, Hawai'i, and nearly every other island across the world's largest ocean to be found and settled—was almost lost forever. Simultaneously, and for shared reasons, the role of *'ulu* and the rest of the canoe crops in the diets of Pacific Islanders declined severely throughout the region. Traditional navigation has experienced a revival since the 1970s, a revival that inspires those working today to restore breadfruit within the diets and landscapes of the Pacific.

When Europeans began to explore the Pacific, their navigators used magnetic compasses to find the cardinal directions, gridded charts to plot their courses, astrolabes and later sextants to determine latitude, and, by the late eighteenth century, marine chronometers to calculate longitude. They found it difficult even to conceive of a navigational system not reliant upon these tools and concepts. The basic problem, though, the key question that needed to be answered, was that which Cook had asked during his final voyage: "How shall we account for this Nation spreading itself so far over this Vast ocean?"[1]

Some Europeans rightly recognized the skills of the Pacific navigators they encountered and understood that their ancestors would have employed these skills to find the islands now inhabited. Johann Reinhold Forster, a naturalist who sailed with Cook, for example, recorded his impressions of "the ingenuity and geographical knowledge" of the Tahitians.

Louis-Antoine de Bougainville, who created the idea of the Pacific islands as Paradise, was so impressed by what he observed in the archipelago that would later become American Samoa that he named an eastern part of the territory Les Isles des Navigateurs, or the Navigators' Islands, for "the skill with which the natives managed their canoes." Another explorer, José de Andía y Varela, wrote that in Tahiti, "there are many sailing-masters among the people" who navigate "with as much precision as the most expert navigator of civilised nations could achieve." Cook himself had previously noted that "these people sail in those Seas from Island to Island for several hundred Leagues, the Sun serving them for a Compass by day, and the Moon and Stars by night," and added presciently his prediction that, "When this comes to be prov'd, we Shall be no longer at a loss to know how the Islands lying in those Seas came to be people'd."[2]

But the voices discounting the Pacific Islanders' Indigenous environmental knowledge and navigational skills—or even denying their existence—very nearly drowned out their praise. Cook later questioned his own convictions and revised his opinion, considering it more likely that the islands had been "people'd" accidentally, by nearshore fishing and trading expeditions being blown off course and the resultant castaways ending up, and remaining to settle, on the remote islands upon which they found themselves. Twentieth-century historian Andrew Sharp was among the most vocal critics of the idea that these island discoveries had been intentional. Sharp wrote, summarizing his view, that "deliberate navigation to and from remote ocean islands was impossible in the days before the plotting of courses with precision instruments," and that "in prehistoric times neither the Polynesians nor any other people in the Pacific performed the feats of deliberate long voyaging and colonization with which they have been credited," concluding that "the detached ocean islands of the world in general, and the Pacific in particular, were settled by accident."[3]

It isn't hard to see why it would appear to early European explorers and modern scholars alike that the Pacific Islanders lacked an ancient, Indigenous navigational system. The introduction of navigational instruments by European mariners, together with changing political and societal factors, turned out to be something of a self-fulfilling prophecy in that the skills and culture of traditional navigation were severely diminished throughout the Pacific during the very early years of outside contact. As European shipping came to dominate the region, and missionaries in several island groups actually forbade long-distance voyaging, the European view of the earth—and the tools and methods used to navigate within it—rose to prominence, and the idea that one would point a canoe toward a star-house and hold course following environmental signs until an island was raised over the horizon quickly fell out of favor.

In 1930 the physicist Paul Dirac published *The Principles of Quantum Mechanics*. He would receive the Nobel Prize three years later for his work on the subject. Among the first concepts Dirac discussed was what has come to be called the "observer effect." Understanding this effect, Dirac wrote, requires "an increasing recognition of the part played by the observer introducing the regularities that appear in the observations." In other words, outsiders viewing a functioning system are likely to affect the very functioning of that system and, therefore, their observations will be not of the system as it was, but of the system as they have affected it. This effect is seen every time you check the air pressure in the tires of your car or bicycle. That hissing sound is air escaping through the valve where you've attached the pressure gauge. The very quantity you're trying to measure is being changed by the act of measurement. As Europeans observed traditional Pacific navigation, they also affected it through their presence and the introduction of new tools, methods, and geographical ideas. What most Europeans observed was an altered system.[4]

But there have always been those who longed to see a pristine system, a pure form of Pacific navigation, existing as it was before being changed. Among the most committed of these purists was David Lewis, a sailor and physician who, in the 1960s, began to visit the few small pockets of the Pacific where traditional navigation methods were still learned and practiced. As a navigator himself, albeit one steeped in the use of modern twentieth-century instruments, Lewis was particularly able to understand the difficulty of sailing far out of sight of land without modern tools. Lewis spent years with master navigators from the Pacific islands, learning and recording their techniques as he literally put them to the test, crisscrossing the ocean, in his words, "entirely without instruments, even clocks . . . full faith being placed in the native navigator."[5]

By the late 1960s and into the 1970s, it appeared to scholars like Lewis that traditional navigation methods might be on the verge of total disappearance. Even in the remote parts of the Pacific where the traditions had held on the longest, they were being replaced by imported techniques and technologies and the remaining navigators were dying of old age.

Lewis, together with anthropologist Ben Finney, believed that traditional navigation in the Pacific could be saved and proposed what Finney called "a nautically appropriate way" to do so: "reconstruct the canoes and ways of navigating, and then test them by sailing between distantly separated islands over which the Polynesians said their ancestors once voyaged."[6]

Finney founded the Polynesian Voyaging Society in 1973 and spent the next two years constructing a voyaging canoe to the best specifications that could be determined from archaeological evidence and historical descriptions and sketches. The vessel was called *Hōkūle'a*, the Hawaiian name for the star known to astronomers as Arcturus, which reaches its zenith directly over the island of Hawai'i and, thus, to a returning Hawaiian voyager, points the way home.[7]

With *Hōkūle'a* ready to sail, the Polynesian Voyaging Society needed a navigator. After considerable searching for someone who knew the old techniques, Mau Piailug, forty-one years old and trained in navigation since infancy, was found on Satawal in the Caroline Islands of Micronesia. Piailug studied the stars of the planned route and, when he was confident, a crew was assembled. In 1976 *Hōkūle'a* made its initial voyage without any navigational instruments from Hawai'i to Tahiti, a distance of more than two thousand nautical miles. Onboard, stocked as provisions, were the familiar Pacific canoe crops, including *'ulu*.[8]

By building a traditional voyaging canoe from ancient and historical plans that had fallen into disuse and entrusting its navigation to the skills and knowledge that had carried the ancient Lapita and Polynesian peoples back and forth across the Pacific, the Polynesian Voyaging Society rescued an entire set of Indigenous nautical traditions from the brink of disappearance. *Hōkūle'a* now sails the world promoting the value of Indigenous knowledge so other traditions might escape the fate that Pacific navigation and voyaging so nearly avoided. Among the traditions most nearly lost following European contact in the Pacific were those related to Indigenous food systems.

Nutritionally, life in the Pacific islands at the time of European contact was usually adequate, sometimes abundant. For example, the Jesuit historian Francisco García wrote that in the Mariana Islands during the early days of Spanish missionary activity, it was "very common to live to be ninety or one hundred years old," citing as an example the observation that "among those who were baptized in the first year of the mission were found more than a hundred and twenty who had passed a hundred years." García attributed this longevity among the CHamoru people to "the robustness of their nature, [and] the uniformity and naturalness of the food without the guile that [we have] introduced."⁹

Elsewhere in the Pacific, good health and nutrition were regularly observed. Food historian Roger Haden recalled how Cook "often reported on the vitality and health of native Pacific Islanders in his journals," specifically noting the "fine teeth" of the Tahitians, observing that the Tongans were "seemingly free from disease," and describing the New Caledonians as "strong, robust, active, well made people." In what would become his namesake archipelago, the Cook Islands, Cook found the population to be "well fed, stout, and active."¹⁰

With such abundance, it did not take long for food-based exchanges to emerge between Europeans and Pacific Islanders. Haden went on to describe the systems that grew out of the provisioning needs of European ships' crews and the "curiosity and covetousness" of local islanders for novel foods and other goods. As the European presence in the Pacific grew, some islanders became increasingly involved in the business of ship provisioning. They would collect fresh produce like breadfruit, coconuts, and yams to trade for European foods and manufactured items. This system of exchanges produced what Haden called "a hybrid food culture" that included both traditional Pacific island foods and those newly introduced from Europe. Among the first culinary introductions were "refined sugar, tea, wheat flour, polished rice, and bully [corned] beef"—none particularly healthy, but not as directly harmful as the alcohol and tobacco that would soon follow.¹¹

Haden presented the history of the Pacific island nation of Kiribati as a typical example for the region. First, in the early nineteenth century, sporadic trading occurred between the local people and visiting whalers. Later, more regular trade links developed, with tobacco and iron being exchanged for locally sourced sea cucumber and turtle shell. Slowly, foreign foods such as wheat flour, tea, canned meat, and tinned fish began to enter the local economy and cuisine. When Kiribati (then called the Gilbert Islands) became a British protectorate in 1892, "local foodways were discouraged," according to Haden, "and Western culture was actively passed on." The convenience and prestige of imported foods, together with the time commitments of wage labor, often led to the abandonment of local food systems in favor of an increasingly foreign diet. The result for islanders' health was not good: according to the World Health Organization, diabetes, heart disease, hypertension, nutrient deficiencies, and obesity have all become prevalent throughout the region.[12]

Another challenge to traditional Pacific foodways was the emergence of monocrop plantation agriculture and its usurpation of land previously used to grow subsistence crops. Copra, sugarcane, and pineapples were some of the earliest cash crops planted on large Pacific plantations. Later, smaller-scale plantations growing citrus, cocoa, coffee, tobacco, and vanilla would emerge. What these crops had in common was that they took up arable land—sometimes an extremely scarce commodity on a small island—that could otherwise have been used to grow food for local consumption.

Cook seems to have foreseen this outcome. During his second circumnavigation, which included stops at New Zealand, Tahiti, and Tasmania, he reflected on the changes he had observed since his first journey of Pacific exploration. "Such are the consequences of a commerce with Europeans," Cook wrote, "We introduce among them wants and perhaps diseases which they never before knew and which serve only to disturb that happy tranquility they and their fore Fathers had injoy'd." Through interaction with foreign sailors, Pacific Islanders developed new desires for new foods, many of which were not only unhealthy but required major shifts in lifestyle to obtain. These lifestyle changes often required landscape changes as well: shifts from subsistence agriculture to commodity crops. The Pacific Islanders' traditional foods had served "their fore Fathers" well; the newly introduced foods failed to do so because that never was their goal. They were, instead, intended to serve global economic markets.[13]

A clear example of the disturbance of the "happy tranquility" Cook had observed in the Pacific can be seen in the case of the *kalu'ulu*, or breadfruit belt of the island of Hawai'i. Before European contact, Indigenous agroforesters in the area around Kona maintained a series of *'ulu* groves twenty miles long and nearly a half-mile wide that, as calculated by environmental scientist Noa Kekuewa Lincoln, produced enough breadfruit to feed between fifty thousand and a hundred thousand people. Under the canopy of their *'ulu* trees, Hawaiians grew bananas, paper mulberry, sweet potatoes, and yams as understory crops. Early visitors to Hawai'i, starting with Cook, recorded their impressions of the organization and productivity of this agricultural system, which they referred to as a series of "plantations"—a term usually reserved for European farming operations. As Hawai'i modernized, however, and as its agriculture shifted to an export-focused model, the breadfruit belt was replaced by a coffee belt. Today, the famous and expensive Kona coffee comes from the exact area that used to provide abundant 'ulu.[14]

But breadfruit and other traditional Pacific foods are coming back, propelled by both a revived interest in healthy eating and a renaissance of cultural and culinary traditions. Noa Kekuewa Lincoln is not only an environmental scientist who studies the kalu'ulu; he's also a cofounder of Māla Kalu'ulu, a cooperative farming operation based upon traditional methods of Hawaiian agroforestry with the goal of restoring a small section of that system on the very same land where it once existed. I visited Māla Kalu'ulu, where Anissa Lucero, the cooperative's vice president, gave me a tour.[15]

What began as a walk through an agroforest, soon turned into a workday when Noa and his partner and Māla Kalu'ulu cofounder, Dana Shapiro, arrived to prune the 'ulu trees. Noa climbed a tree with a chainsaw in hand and Anissa and I each grabbed a pair of loppers to cut smaller branches. Māla Kalu'ulu practices "chop and drop," using a tree's fallen leaves and other cuttings as mulch to hold moisture and provide nutrients at the base of the trunk. The farm's stated mission is to "learn from the past to create a more resilient food future." In that future, Noa, Dana, and Anissa believe, people in Hawai'i are going to be eating more 'ulu.

In being resurrected, traditional Pacific foods like 'ulu are sometimes changed, sometimes given a modern twist, often incorporating elements of cultures that have overlain those of the Indigenous Pacific Islanders. The difference is that these newer cultural arrivals are no longer entirely replacing or obscuring traditions. Instead, the traditions are rising up through the layers to reemerge, sometimes having picked up new elements along the way.

For example, on a single day in Tahiti, I saw two clear examples of breadfruit's modern, blended representation. Upon entering the Notre-Dame cathedral in Pape'ete, I was greeted by a statue, carved from wood, depicting the Madonna and Child. Baby Jesus clutches a large, round, mature breadfruit—an element of the Advent story absent both from the Gospels and from most nativity scenes I have encountered. Leaving the church, I took a short walk down a street named for Charles de Gaulle and stopped at a sidewalk café called Maru Maru for dinner. Here I ordered steamed parrotfish with gnocchi de 'uru, dumplings inspired by the Italian potato dish but with breadfruit in the primary starchy role.[16]

Traveling around the Pacific, I saw example upon example of breadfruit's reemergence. Some, like the church statue and the gnocchi,

Figure 5.1 A statue of Mary holding baby Jesus, holding a breadfruit, at Cathédrale Notre-Dame de Pape'ete, Tahiti, French Polynesia. (Photograph by Russell Fielding.)

represented insertions of breadfruit into imported traditions—Christianity and pasta. Often these "fusion" efforts were led by people who themselves were newcomers to the Pacific. Other times I witnessed more overt forms of reclamation, led by Pacific Islanders seeking to reestablish breadfruit in its traditional roles within Pacific societies.

In Sāmoa I met Seeseei Molimau-Samasoni, a researcher at the Scientific Research Organisation of Sāmoa (SROS), which operates similarly to an agricultural extension office in the United States, providing free consultations to Samoan farmers. When I arrived during a downpour at the end of SROS's long, breadfruit tree-lined driveway, Seeseei was there to greet me: "You've come to the home of *ulu*!" she announced from beneath her umbrella. Seeseei explained to me that in Sāmoa, as in many other Pacific island nations, traditional starches like *'ulu* have lost popularity to imports. Even just considering fruits, people prefer imports to traditional canoe crops. If Samoans can't afford imported apples and oranges, she said, many simply don't eat fruit, choosing to ignore the bounty growing on their own local trees. SROS is trying to change that. In 2017 the organization hosted a global breadfruit summit, in which experts from around the world came together to discuss growing and promoting breadfruit. Seeseei's organization sponsors research into breadfruit processing, extending the postharvest shelf life, and the creation of new breadfruit-based products including both foods and cosmetics. They also study people's preferences and work with chefs to come up with new recipes that might become popular enough to compete with imported foods.[17]

A big promoter of these modern *'ulu* recipes is Sam Choy, a chef whose 2022 *'Ulu Cookbook* contains more than one hundred breadfruit recipes, comprising a cuisine Choy has called "Hawaiian-fusion." Choy is a culinary celebrity in Hawai'i, having hosted his own locally broadcast cooking show, *Sam Choy's Kitchen*, which was followed by the similarly titled *Sam Choy's in the Kitchen*. In the latter program, Choy visits families in their Hawaiian homes with the goal of creating delicious, spontaneous meals from whatever foods are already available in their pantries and refrigerators. Watching old episodes online, you can see Sam's imagination light up when he finds something like "a Ziploc bag of roasted *'ulu* chunks" as one of the available ingredients.[18]

Other celebrity chefs have been more confounded by breadfruit. When Gordon Ramsay visited Maui in 2019 as part of his *Uncharted* series, he wondered, "How can I turn this thing into something delicious?" admitting that ʻulu was "not the most appetizing-looking fruit-slash-vegetable." Still, in the manufactured urgency of a reality-television challenge, Ramsay was able to create innovative dishes with traditional Hawaiian ingredients. Mashed breadfruit stood in for potatoes in a shepherd's pie that certainly would have met Sam Choy's definition of Hawaiian-fusion. Instead of lamb, the pie included meat from the axis deer, a nonnative species introduced to Hawaiʻi during the nineteenth century. Without the predators of their natural Asian habitat, axis deer populations have grown large, and hunting is promoted as a method of population control. For dessert, Ramsay served a taro-based custard—which he named "*poi*-nacotta"—in coconut half-shells. All were well received by his assembled guests, with the ʻulu-based Hawaiian shepherd's pie receiving special compliments.[19]

In Honolulu, wanting to see this process of culinary fusion myself, I met up with Sarah Burchard and Harrison Ines, a pair of culinary innovators who work to find new ways of preparing ʻulu with the goal of expanding its role in contemporary Hawaiian food systems. Over the years, through various food-related research, I have learned that when you study fisheries you go fishing, when you survey farms you return home with bushels of produce, and when you interview chefs you cook.

Sarah, Harrison, and I discussed food traditions both old and new while working together in their apartment's tiny, fifty-square-foot kitchen to produce an ʻulu-based meal. Harrison chopped steamed ʻulu quarters for a vegetarian chowder and Sarah swiped a few slices for a warm ʻulu salad. I mixed ʻulu flour with the other ingredients needed for fried-chicken batter. As we cooked, and then as we ate, we talked about how profoundly Hawaiʻi has changed, both ecologically and socially, since colonization, and we noted the irony of our group—two-thirds *haole*, as white people in Hawaiʻi are called, with only one Hawaiʻi-born *kamaʻāina*, himself of Filipino descent—producing a Hawaiian-fusion meal that, while it incorporated plenty of locally grown ʻulu, drew significant influence from the cuisines of New England and the southern United States.[20]

Sarah and Harrison serve as ʻUlu Ambassadors for the Hawaiʻi ʻUlu Cooperative. Formed in 2016, the cooperative is, according to its website, "a farmer-owned business working to revitalize ʻulu . . . by empowering farmers as change-makers in Hawaiʻi's food system." In the years since its establishment, the cooperative has expanded beyond empowering farmers alone. Its website, eatbreadfruit.com, lists more than a hundred ʻulu-based recipes. Many involve replacing potatoes with breadfruit: mashed ʻulu, scalloped ʻulu, ʻulu beef stew, ʻulu latkes. When peeling a raw breadfruit for one of these recipes, don't toss the skin into your compost! If the fruit was washed well first, strips of ʻulu skin can be pan-fried in butter or oil, then lightly seasoned with salt and pepper to make a delicious version of hashbrowns. My children say this is their favorite breadfruit recipe.[21]

The Hawaiʻi ʻUlu Cooperative is also in the business of breadfruit processing and sales. At their collection and production facility near Kona, I met Elias Ednie, who offered to give me a tour. Eli, as he's called, explained that the cooperative buys ʻulu from any grower but gives preferential prices to its own members. Anyone in Hawaiʻi with at least one ʻulu tree can become a member, after being approved by the cooperative's board and paying a startup fee. On Mondays and Tuesdays, the cooperative buys ʻulu—fifty pounds minimum—from anyone who brings it in. They have a satellite collection facility in Hilo, on the other side of the island, as well.

Eli showed me the cooperative's equipment. There's a large trough with water jets for washing, a stainless-steel peeling machine with its buttons labeled in both Japanese and English that Eli described as "very fast," a steam oven, a flour mill taller than I am, and a walk-in freezer. The cooperative's two main products are steamed and frozen ʻulu quarters and ʻulu flour, but they also sell hummus and chocolate mousse made from ʻulu, as well as steamed and frozen squash, sweet potato, and taro. Retail customers can buy the cooperative's products online via their website or at shops throughout Hawaiʻi. The online product locator map lists well over a hundred outlets spread across the islands of Hawaiʻi, Maui, Oʻahu, and Kauaʻi—and even one each on the small islands of Lānaʻi and Molokaʻi.

After visiting with Eli, I drove around the island of Hawaiʻi for several days, meeting the cooperative's member farmers and learning about their experiences. Jack Turner of Captain's Ohana Farm in Hōnaunau reminisced about the time he "couldn't give away" all the breadfruit that his nineteen trees produced. He was just about to cut them all down to make room for more coffee—the farm's main crop—when he overheard someone at a bar in Kona talking about the cooperative. Now Jack is a member, and his ʻulu trees still stand.

Far up north on the Kohala Peninsula, I jogged around the One Village Farm, trying to keep up with farm manager Travis Dodson as he zoomed about on a kind of one-wheeled, off-road, electric skateboard. Whenever he stopped, Travis would explain something new: the principles of syntropic agriculture or the way the ancient Hawaiian canoe builders managed ʻulu groves to produce tall, straight trees to be harvested for their lumber. We snacked on rollinia, a delicious fruit imported to Hawai'i from South America, which now thrives on the One Village Farm. And, like most farmers I met, Travis gave me armloads of *'ulu* to bring back and cook.

One of the most interesting meetings I had with a cooperative member was when I met Duane Lammers at Hana Ranch on Maui. When my family and I arrived, Duane stepped out of the ranch's main office building wearing polished boots, Wrangler jeans, a saucer-sized belt-buckle, a white button-up shirt, and a big white hat. "Daddy, that's a cowboy," my daughter, then four, whispered. Indeed, before coming to Maui to manage Hana Ranch, Duane had raised bison in South Dakota. Now he serves on the Hawai'i 'Ulu Cooperative's board of directors and grows breadfruit on the three thousand acres under his management while cattle graze in the trees' shade. When we met, the ranch supported just over one hundred trees. Duane's goal was two thousand.

"*'Ulu*, to me, is like bison," Duane told me, and paused as I tried to think of *any* way that a breadfruit tree might be compared to a buffalo. "Both are easy to take care of," he continued. Another similarity, according to Duane, was the market. "In the 1970s, people thought bison were going extinct," and they nearly did. When ranchers started increasing their herds and bison meat began to appear in North American grocery stores, it was obscure, but those in the know viewed it as a healthier alternative to beef. Now, Duane says of *'ulu*, "hardly anybody knows anything about it, but those that do have only positive things to say." His bison analogy came into focus. If consumers, particularly outside of the Pacific islands, were simply aware of breadfruit, it might be seen as a healthier, more sustainable alternative to potatoes, rice, and other common starches. *'Ulu* flour might replace wheat for some consumers, particularly those who, by choice or necessity, follow a gluten-free diet. "My gut just tells me this is a crop that's gonna go somewhere," Duane told me. And I felt that when a cowboy speaks from his gut, I should listen.

Everything I had learned about new ways to grow, promote, sell, and serve breadfruit in Hawai'i and the rest of the Pacific became especially poignant when I met someone trying hard to keep the old connections among people, food, and the land alive. Tammy Mahealani Smith is the director of culinary services at Lunalilo Home, a residential nursing facility in eastern O'ahu. I met Aunty Tammy, as she's called, in the garden that she planted for the home's elderly residents to enjoy traditional Hawaiian food, grown on site. Lunalilo Home was established in 1883 as part of the estate of King Lunalilo, Hawai'i's first popularly elected monarch, whose will set aside funds for "the erection of a building or buildings on the Island of O'ahu, of iron, stone, brick or other fire-proof material, for the use and accommodation of poor, destitute and infirm people of Hawaiian (aboriginal) blood or extraction, giving preference to old people."[22]

As is common for many places in Hawai'i though, the buildings are not what truly make the place, no matter how solidly built or fireproof they may be. Rather, it is the plants. The grounds at Lunalilo Home feature networks of walking paths lined with bougainvillea, heliconia, hibiscus, 'ōhi'a, plumeria, and various other flowering plants. In addition to the cultivation of these ornamentals, Aunty Tammy has overseen the development of a native food garden. The king's original focus on people of aboriginal blood has been expanded; Lunalilo Home now serves "kūpuna [elders] of all ethnicities," according to its website. Residents who are mobile can enjoy the experience of walking or traveling by wheelchair outdoors among taro plants, bananas, and 'ulu trees. Many of the residents in long-term care at Lunalilo Home suffer from dementia, but, Tammy explained, some experience moments of clarity when the scents of food from their childhoods waft into the dining room from the kitchen.[23]

Studies have revealed the power of both food and music in promoting the recall of memories among people experiencing dementia. Lunalilo Home uses both, with Tammy's staff preparing nutritious, culturally relevant meals from locally grown ingredients and local musicians dropping by for weekly *kanikapila*, or informal jam sessions. Diane Paloma, former CEO of the Lunalilo Trust explained to a local journalist in 2020, referring to the home's residents, "You see, here at Lunalilo, we're not just trying to feed their physical bodies, we're trying to feed their souls."[24]

In addition to its contributions to the mental, physical, and spiritual health of the residents, the Hawaiian garden at Lunalilo Home also provides food security in times of crisis. Tammy told me that, when COVID-19 hit, "we were ready." The local systems she had established to provide O'ahu-grown food for her residents, including their own garden, weren't nearly as affected by the pandemic as were larger-scale food-importation networks. In Hawai'i you often hear that 80 to 90 percent of the food is imported, mostly shipped from the mainland United States. Whether due to a pandemic, a natural disaster, or some other interruption, local food production systems can provide security when global systems fail.[25]

Tammy also sees her work with ʻulu as part of her kuleana, her reciprocal rights and responsibilities, to keep Hawaiian food on Hawaiian plates. She told me the story of the "TMT protests," an environmentalist and Indigenous-rights movement that seeks, specifically, to halt construction of the Thirty Meter Telescope atop Mauna Kea—the highest point in the state of Hawaiʻi and a place of sacred importance to Hawaiian people—and generally to oppose the ongoing colonization of Hawaiʻi following the 1893 overthrow of the Hawaiian monarchy. While telescopes had already been constructed atop Mauna Kea and on the summit of Maui's Haleakalā, the proposed TMT, to some, seemed a step too far toward the industrialization of these sacred and ecologically fragile mountains. The protests began in 2014, after the Hawaiʻi Board of Land and Natural Resources had approved the TMT project the previous year.[26]

In planning a march down Waikīkī's Kalākaua Avenue, protest organizers contacted Aunty Tammy, asking if she would oversee the preparation of lunch for several hundred demonstrators. "Of course I will; what would you like me to make?" she asked. The organizers suggested hotdogs. Considering the very purpose of the event, the assertion of Hawaiian agency over Hawaiian land, Tammy gently made a countersuggestion: "maybe we should serve something Hawaiian." While in the end, the participants in the march ate traditional Hawaiian laulau (ʻulu and taro wrapped in luʻau and tī leaves), the original suggestion might imply just how pervasive the colonial mindset has become, even in overtly anticolonial Hawaiian thought.

The memory of the protest march seemed to serve as a trigger. "We are fighting for space," Tammy told me, her eyes—all I could see of her face above her KN95 mask—welling up with tears. This was the summer of 2021; we were both wearing masks, and social distancing protocols were strict, especially in a place like Lunalilo Home with so many vulnerable residents. Still, feeling that compassion was at least as important as public health, I put my arm around Aunty Tammy as she laid her head on my shoulder and cried. I remembered hearing Israel Kamakawiwoʻole's sweet tenor voice singing "Cry for the gods, cry for the people / Cry for the land that was taken away" as Tammy wept over the history of her islands, their people and their produce. "Mainlanders grow more ʻulu than Hawaiians," she sobbed, referring to people who had moved to Hawaiʻi, bought land, and begun farming traditional Hawaiian crops like breadfruit, often as a hobby, like many of the ʻulu growers I had met. "They're trying to claim it," Tammy insisted. When I asked what for, a one-word answer was sufficient: "Profit."[27]

Tammy doesn't exactly begrudge the mainlanders—though she did state her opinion that the Hawai'i 'Ulu Cooperative includes too great a ratio of *haole* to Native Hawaiian farmers. She just wishes more Hawaiians would, in her words, "get educated and get dirty" by returning to the land. I mentioned Michael Pollan, the author known for his food-focused writing, who cited a famous line from the farmer, poet, and essayist Wendell Berry: "eating is an agricultural act." "He might have added that it's a political act as well," Pollan commented. Aunty Tammy wholeheartedly agreed that eating, especially in Hawai'i, is an act both agricultural and political.[28]

In 1893 Queen Lili'uokalani, monarch of the Kingdom of Hawai'i, gave up her throne under the threat of violence that seemed inevitable as a coalition of American and European businessmen, planters, and missionaries initiated the overthrow of her kingdom. Lili'uokalani's official, written response to being deposed stated that "to avoid any collision of armed forces, and perhaps the loss of life, I do, under this protest and impelled by said forces, yield my authority until such time as the Government of the United States shall, upon the facts being presented to it, undo the action of its representative and reinstate me in the authority which I claim as the constitutional sovereign of the Hawaiian Islands." Of course, the United States never undid any such thing. Hawai'i was annexed as a U.S. territory five years later. Statehood would follow in 1959.[29]

At the ceremony marking the transition of power, the Royal Hawaiian Band played "Hawai'i Pono'ī," the national anthem of the Kingdom of Hawai'i, now Hawai'i's state song. As the Hawaiian flag was lowered and the American flag raised, some members of the band refused to play "The Star-Spangled Banner" or to sign the oath of allegiance to the newly formed provisional government. They were threatened for their disloyalty, warned that they would lose their jobs, would be shunned from the new society being formed, and would be left with nothing but rocks to eat. The band members are said to have responded, *"Ua lawa mākou i ka pōhaku."* We are satisfied with rocks.

In the aftermath of the overthrow, a song that Queen Lili'uokalani had written several years earlier began to take on new meaning. "Aloha 'oe," usually translated "Farewell to Thee," is said to have been inspired originally by the sight of a good-bye embrace between lovers. But the loss of her kingdom seemed to lend a preternatural relevance to the queen's mournful, melancholic tune, the slow, sad lyrics, accompanied by the iconic Hawaiian slack-key guitar. Today, when I hear "Aloha 'oe," I imagine that last Hawaiian monarch bidding farewell to her kingdom and offering her land, her people,

> One fond embrace
> Ere I depart
> Until we meet again.[30]

From these events another song emerged, written by Ellen Keko'aohiwaikalani Wright Prendergast, a personal friend and confidante of Queen Lili'uokalani, but one who had opposed the queen's acquiescence. Prendergast had offered quarter in her home to Royal Hawaiian Band members being punished for their disloyalty. These musicians had undoubtedly told her their story, which inspired her to write a song called "Mele 'Ai Pōhaku" (The Stone-Eating Song). Prendergast's lyrics are defiant where "Aloha 'oe" had been mournful, promising resistance where the queen had offered a final fond embrace. Prendergast wrote,

> 'A'ole mākou a'e minamina,
> I ka pu'ukālā o ke aupuni,
> Ua lawa mākou i ka pōhaku,
> I ka 'ai kamaha'o ka 'āina,

which can be translated,

> We do not value
> The government's hills of money,
> We are satisfied with rocks,
> The wondrous food of the land.

In other words, regardless of changes both cultural and political, the 'āina, the land of Hawai'i—and, by extension, the rest of the Pacific islands—would always provide bountiful and sufficient food for its people.[31]

On the island of Guam, I got to see perhaps the clearest example of "the wondrous food of the land" on a trip with Bob Bevacqua, professor of horticulture at the University of Guam, and his graduate student, Ajalyn

Omelau, to the War in the Pacific National Historical Park at Asan Beach. This is the site of the U.S. Marines' invasion on July 21, 1944, which initiated the ultimately successful effort to retake Guam from the Japanese during the Second World War. This 1.5-mile-long, north-facing beach is backed by a large grove of tall coconut palms—one tree for each marine killed in the invasion, Bob explained. At the beach's western limit, a peninsula reaches into the sea to form Asan Point and extends inland as Asan Ridge. Along the ridge, more than two hundred breadfruit trees of Guam's endemic 'Lemmai' cultivar grow. Every summer, when the limbs of these trees hang heavy with breadfruit, the public are invited to come and harvest for their own consumption.

Bob described to me the sight of dozens of CHamoru children climbing high and returning to their waiting families below with arms cradling breadfruit. He confessed worrying about their risk of falling from so great a height; unmanaged, these trees grow far taller than a carefully pruned backyard breadfruit tree. Most families take only what they can consume in a day or two, but some see a business opportunity. Pickup trucks are occasionally seen leaving Asan Bay with beds full of hastily piled breadfruit, often partially concealed beneath tarpaulins. The problem for these would-be entrepreneurs, according to Bob, is that the market is so glutted with breadfruit at harvest time that no one is buying, especially since all you have to do is drive down to Asan Bay and pick your own for free.

In the Bible, the book of Exodus relates the story of "bread from heaven," known as manna, which God provided daily to the desert-wandering Israelite people, along with the instruction to gather only enough for each day. Those who disregarded this rule and took extra manna to last until the next day or beyond would awake to find that in the morning "it bred worms, and stank.... and when the sun waxed hot, it melted." Likewise, those takers of too much breadfruit at Asan Bay watch as their supply quickly browns and rots, wasting away in a market with no buyers: "our daily bread," quite literally.[32]

The history of 'ulu in the Pacific is one of a glorious past, a period of neglect, and a contemporary renaissance. As part of the suite of canoe crops, 'ulu was instrumental in the original Lapita and Polynesian navigators' extraordinary feats of sailing, landfinding, and settlement. Early Pacific Islanders discovered myriad uses for the products of the breadfruit tree including food, medicine, building materials, and clothing. Its preservability was insurance against famine. Its abundance allowed the development of complex civilizations and structures. As a literal enactment of the Hawaiian proverb *"He keiki aloha nā mea kanu"* (Beloved children are the plants), *'ulu* saplings are still planted above the buried placenta or umbilical cords of newborn babies in the Pacific today. The trees that grow will both symbolize and provide materially for each child as she grows.

When Europeans came to the Pacific, their thoughts on breadfruit were conflicted. This unfamiliar crop appeared to produce abundant food while requiring minimal effort. It seemed to contradict the biblical notion that people must earn their food through labor and, to some, hinted that the Pacific Islanders may have remained innocent of the original sin that brought this curse upon mankind. This belief, along with the physical beauty of Pacific island landscapes and their inhabitants, provoked admiration among some Europeans, disdain among others. At the center of these conflicted feelings, some Europeans considered how to "mend" the Pacific Islanders of their ways, largely by introducing a Western work ethic and, in some cases, by cutting down breadfruit trees. Others saw in breadfruit a solution to food shortages and began to think of bringing it to other islands as a fuel for enslaved workforces.

Over time, it was not the zealous destruction of breadfruit trees that changed the foodways of the Pacific islands, but colonization—cultural, economic, and political—and the gradual replacement of traditional foods with imports from a globalized market. Canoe crops fell out of favor as commodities like coffee, pineapples, and sugar supplanted the *'ulu* groves and taro patches of the islands' agricultural landscapes and imported starches like rice claimed a greater role in islanders' diets. Recently though, a renaissance of local and traditional foods has begun as part of larger cultural sovereignty movements throughout the Pacific, and thanks to the dedicated work of a diverse cadre of chefs, farmers, producers, and scientists, *'ulu* is being reelevated toward its former glory.

PART II

The Caribbean

Map 2 The Caribbean region. (Cartography by Alison DeGraff Ollivierre, Tombolo Maps & Design.)

CHAPTER VI

A Dish de Résistance

Breadfruit is being redeemed. During the last decade of the eighteenth century the British brought hundreds of breadfruit saplings from islands in the Pacific to the sugar plantations of the Caribbean. They did so based upon reports that it was productive and required little tending; it could save effort, allowing the people they had enslaved to spend more time working the sugarcane while their food essentially grew itself. The British acquired root shoots of breadfruit from Tahiti in a fair exchange; they took people from Africa against their will. A deracinated human population, forced to labor without pay and without end, was given as food the fruit of a literally uprooted tree. Literary scholar Hannah Rachel Cole has referred to breadfruit in the Caribbean as "a vegetal legacy of enslavement. . . . a botanical projection of colonial power."[1]

But just as they claimed their freedom and took ownership of their islands, the Caribbean people reclaimed—and are still reclaiming—breadfruit. It now adorns the Caribbean table alongside other imported foods: the ackee, callaloo, mango, and plantain of modern West Indian cuisine. Yes, it's a mélange of flavors and traditions revealing the African, Asian, and European origins of Caribbean cooking. Even some Indigenous influences can still be found: travel deeply enough in the region and you'll come across menus featuring iguana and manicou. All those ingredients have been well stirred and set to simmer. What you smell when you lift the lid, and taste when, finally, dinner is served, is something local and new, creolized to a point greater and more representative than the sum of its parts. Caribbean cuisine is no longer just a mix of other cultures; it is its own—a multicultural communion developed anew within this "rosary of archipelagoes." Breadfruit, today, is becoming as native to the Caribbean as *'ulu* is to the Pacific.[2]

Historically speaking, this is a recent development. For much of its history in the region, breadfruit was distrusted by many, even despised by some. V. S. Naipaul, of Trinidad, derisively called breadfruit "cheap slave food" in his classic travelogue *The Middle Passage*. Antigua-born author Jamaica Kincaid wrote about "the much hated breadfruit," and as recently as 1999 she could still say that "no West Indian that I know has ever liked it." Breadfruit, now an ingredient in the national dishes of at least three Caribbean countries (oil-down in Grenada; stewed saltfish, spicy plantains, seasoned breadfruit, and coconut dumplings in St. Kitts and Nevis; and fried jackfish with roast breadfruit in St. Vincent and the Grenadines), is being reclaimed like the islands themselves. Linguist Michel Erman has written that "a national dish is a dish *de résistance*." Resistance of colonial rule is seen in both the cuisines and the political histories of the Caribbean. Beginning with Haiti in 1804, formerly enslaved people and their descendants have established independent countries where once cash-crop plantation colonies stood. Some of the larger islands gained their independence through war: Toussaint L'Ouverture led Haiti's bloody revolution against France and Cuba's *hombre sincero*, José Martí, set his country on its violent course toward freedom from Spain. But most of the Caribbean's now-independent islands went peacefully.[3]

Caribbean schoolchildren learn of one of Barbados's national heroes, Errol Barrow, who "fought with words," that is to say, not with weapons, against Britain to gain his country's independence nonviolently in 1966. Trinidad and Tobago had declared independence four years earlier. Grenada, Dominica, St. Lucia, and St. Vincent and the Grenadines followed during the 1970s; the two-island nations of Antigua and Barbuda and St. Kitts and Nevis went soon after. The remaining colonized islands—Anguilla, the Caymans, Montserrat, Puerto Rico, the Turks and Caicos, and both sets of Virgin Islands, along with most of the French and Dutch islands—have mainly opted to retain connections to their metropoles. Politically, the region went through major changes during the twentieth century. Most of the islands are now postcolonial. But, as historical geographer David Lowenthal has advised his readers, "in the Caribbean the past is a living presence." Every breadfruit tree rooted in soil between the Florida Keys and the Orinoco Delta contributes to the living presence of the Caribbean's past, a past defined by abuse, plunder, replacement, and—eventually—the reclamation of value carried over from a colonial world.[4]

It is through that reclamation that breadfruit today has earned its status as a dish de résistance across the Caribbean. When a new country emerges as independent, or when an enslaved population gains emancipation, the newly free people can choose which legacies of their former rulers they discard and which they claim as their own. Those they keep, they often reinvent; consider the English-based patois you hear when traveling through Jamaica, the calypso-tinged Charles Wesley hymns I heard sung at the Brighton Methodist Church in St. Vincent, and, in the case of breadfruit, a "cheap slave food" turned into a national dish. This transformation is not without precedent. Breadfruit had been part of an infamous act of resistance once before.

Today the international shipment of plant materials is highly regulated. Without proper permits in place, plants, seeds, and fruits usually cannot be carried from one country to another, and even then, there are often compulsory periods of quarantine and various pesticide treatments required. I once made the mistake of arriving on a flight to Auckland with luggage that contained my old hiking boots, unwashed since my most recent jaunt down some Appalachian trail. The biosecurity officer whose clearance I was required to obtain before entering New Zealand, looked at me mournfully as I placed my mud-caked Vasques onto a stainless-steel countertop. He disappeared with the boots into a back room. I remembered reading about a celebrity—I'd forgotten who—who had recently been fined hundreds of dollars for failing to declare an apple that she had in her carry-on bag, and I worried how much I might have to pay for my inadvertent smuggling of who-knows-what North American soils and seeds. Imagine my relief, then, when the agent returned, my freshly scrubbed boots in hand, saying, "no worries, mate; next time just give 'em a better rinse." The kind of biological transfers that guardians of island ecosystems—like my new Kiwi best friend—assiduously try to avoid today, were once celebrated as triumphs of "economic botany."[5]

Historian Alfred Crosby coined the term "Columbian exchange" to describe the massive increase in the intercontinental movements of plants, animals, people, diseases, and ideas following Christopher Columbus's first transatlantic voyage in 1492. As European governments steadily increased their colonial holdings throughout the Americas and beyond, their naturalists sought to pair newly "discovered" plants with botanical or agricultural niches, then transplanted seeds, roots, or cuttings to new soils where they might provide foods, medicines, commodities, or luxuries for cultivation and consumption by new peoples. Economic opportunities for Europeans were immense, with new products being brought to both European and colonial markets, and plantations being established to produce new commodities at greater scale. As the colonial reach of Europe's empires expanded into the Pacific, so did the travels of so-called bioprospectors, whose efforts to find new plants echoed earlier explorations of the Americas by conquistadors in search of gold and other riches. Compared to the exhaustible mineral wealth sought by the conquistadors, the renewable nature of the "rich vegetable organisms"—what historian Londa Schiebinger has called "green gold"—being found throughout the tropics made these plants even more valuable to the Europeans who would transplant, cultivate, and profit from them.[6]

When the mission to introduce breadfruit into the Caribbean from the Pacific was first conceived, Europeans had already nearly completed their replacement of the Caribbean islands' Indigenous human populations with enslaved laborers taken from Africa. They had replaced the islands' native forests with ordered fields of sugarcane imported from Brazil but first domesticated in Southeast Asia. They had even imported vervet monkeys from Africa, a ubiquitous scourge to agriculture now on some islands, apparently on a whim and—if the seventeenth- and eighteenth-century Dominican missionary Père Labat is to be believed—at a scale two orders of magnitude greater than intended, owing to a simple miscommunication.[7]

Following the introductions of enslaved people from Africa, cash crops from Asia, and crop-raiding monkeys by mistake, who could argue with a plan to import a fruit-bearing tree from the far side of the world? Especially if this tree really was as productive as its reputation would suggest. Perhaps, with the addition of the breadfruit, the Caribbean islands might attain a degree of the utopian, paradisical qualities associated then with their Pacific counterparts.

By the middle of the eighteenth century, Europe's colonialist countries had established systems of what environmental historian Richard Grove has called "green imperialism," built upon parallel networks of linked botanical gardens spread throughout the tropical world. Like the twin foci of an ellipse, French efforts were centered upon Mauritius's Jardin Botanique des Pamplemousses and the Jardin des Plantes in Paris. The British had a more extensive network of tropical gardens. Grove has described a worldwide system, beginning in about 1770, by which more than a hundred "collectors" would range throughout the tropical locales of the British Empire, providing rare and unknown (to the British) plants for about fifteen gardens established mainly in South Asia and the Caribbean. From Kew Gardens outside London, Joseph Banks oversaw the entire operation. Once new specimens were established in a garden, they would be transplanted elsewhere as part of what historian Toby Musgrave has referred to as reciprocation "in floral kind," first to Kew where they were unlikely to grow outside of greenhouses, but could be rigorously examined and formally described, and then from garden to garden across the tropics. Grove described "the main tropical axis" for these intergarden transfers as "running between Calcutta and St. Vincent and having a central and essential transit point at St. Helena." Indeed, when breadfruit was finally, successfully brought from the Pacific to the Caribbean, it first stopped at the remote Atlantic island of St. Helena—Napoleon Bonaparte's later place of exile—where twelve young trees were delivered before the rest were to be carried across to St. Vincent and then Jamaica.[8]

Anyone who has visited a botanical garden in the tropics understands the powerful sense of calm that such a landscape produces. Historian Richard Drayton channeled Enlightenment sensibilities and spiritualities when he referred to the botanical garden as "the means through which the power of God might both be understood and mustered to human purposes, an Eden in the fallen world." The older gardens, in particular, tend to be located near urban centers, which have often grown up to surround the gardens, creating oases in the midst of bustling cities. Such is the case for the Acharya Jagadish Chandra Bose Indian Botanic Garden, established in 1786 as The Honourable Company's Botanic Garden, Calcutta. As Calcutta grew from a company town of just over a hundred thousand inhabitants to the modern megacity of Kolkata—the name was changed in 2001—the contrast between its densifying city streets and the winding paths of the garden became even more pronounced. People retreated to the garden, and still do, as respite from urban oppression. As early as 1791, Banks found it necessary to remind the East India Company leadership that the garden had been established "for utility and for science and not as a place of retreat for officials."[9]

The same is true of the St. Vincent Botanical Gardens, established in 1765 on the outskirts of the country's capital, Kingstown. Today, as the bustling city has crept uphill from the harbor to surround the garden, its calming and oasitic nature has only increased. During my long periods of research fieldwork in St. Vincent, I would often use the Gardens "as a place of retreat," exactly as Banks had instructed the East India Company's officials *not* to do. Each time I visited, I would make the long circuit to the back of the property to visit the Garden's—and the island's—most famous breadfruit tree. Near the back fence, this venerable tree stands tall; unpruned, its lowest branches are far out of reach. A small sign placed at the tree's base identifies the species and notes that this tree was grown from "a sucker from one of the original plants introduced by Captain Bligh in 1793." While this claim to provenance is likely true, it's also not very unique. Because most breadfruit cultivars do not produce seeds, and because even those that do are normally not grown from seeds, it's likely that most breadfruit trees on St. Vincent derive from air-layers or root shoots taken from the trees originally introduced to the Caribbean, or those trees' own generational descendants.

The success of the St. Vincent Botanical Gardens aroused some degree of envy among eighteenth-century planters in Jamaica, who soon lobbied the colonial government for similar gardens to be established on their island. These planters would have been loath to give up arable land for anything other than sugar, except when the benefits outweighed the costs. As such, the gardens in both St. Vincent and Jamaica were explicitly established as part of the worldwide network of plant transfers, which in turn directly benefitted the Caribbean planters. Plants found by collectors in one part of the world that the scientific staff at Kew Gardens assayed to be economically valuable in another would be transferred—often at great effort and expense—and tested first in botanical gardens like those in Calcutta and St. Vincent. Those that survived and produced the desired food, fiber, or medicinal crops in sufficient quantities while planted in their new soil, would be propagated and distributed to local plantations.

By the 1780s, the British government had resolved to send a ship to the Pacific for breadfruit. The trees were to be transferred to the botanical gardens on St. Vincent and Jamaica, from which, after observation and testing to ensure that they could thrive in this new environment, they would be distributed among the sugar-producing islands of the Caribbean. Jamaica Kincaid has summarized the reasoning succinctly: "Those many slaves had to eat food. And they had to have time to cultivate it. It isn't hard to imagine the calculation: If an easy source of nourishment could be found for these people, they would produce more valuable goods instead of growing their own food." The Royal Navy purchased a used collier named *Bethia* and rechristened it something more fitting for its upcoming mission: they called the ship HMS *Bounty*.[10]

William Bligh was appointed commander of the ship, an appropriate choice for a voyage planned for both the Pacific and the Caribbean: Bligh had sailed with Cook aboard HMS *Resolution*, and even served as the de facto commander of that ship after Cook was killed in Hawai'i. He had also captained merchant vessels sailing regularly between England and the Caribbean, so he was familiar with both endpoints of the proposed breadfruit voyage. Two botanists were selected to serve as gardeners, caring for the potted plants on what was anticipated to be their long hemispheric journey. In addition to Bligh and the gardeners, the crew comprised forty-three sailors from the Royal Navy.[11]

After ten months' sailing, the *Bounty* reached Tahiti in October 1788. The sailors were understandably relieved to see land, especially such a lovely and storied island as this. The *Bounty* anchored in Matavai Bay, a partially sheltered cove on Tahiti's north shore. Here most of the crew spent more than five months living on shore, cutting and potting the root shoots of breadfruit trees in collaboration with local Tahitians, and, as Bligh would later write, making "some female connections." Historian Richard Hough later clarified Bligh's oblique reference to "the promiscuity that began within minutes of the *Bounty* dropping anchor at Matavai." Some of the relationships that developed appear to have been more committed than mere fleeting promiscuities. Fletcher Christian, who had signed on as master's mate and was later promoted to acting lieutenant, for example, established a lasting relationship with a woman named Mauatua, whom he would later marry.[12]

By April 1789, more than a thousand potted breadfruit saplings had been loaded onto the *Bounty*'s decks and into former cabins, since converted to greenhouses. The sailors were back on board, some against their will, and the *Bounty* began its long journey toward the Caribbean, the first part of its mission, to acquire breadfruit, accomplished and the second part, to deliver breadfruit, now underway. Bligh had always been known as a stern commander, but during the first two and a half weeks out of Tahiti he was given to exceptionally "violent outbreaks against his officers in general, and

Fletcher Christian in particular," according to Hough. While he was certainly not averse to the lash, Bligh's most notorious abuses were verbal, not physical. This mistreatment, along with the prospect of many months at sea and an eventual return to England, was especially hard for the men to bear when contrasted with the relatively idle and idyllic life they had lived on Tahiti, and the affection they had experienced with the Tahitian women.[13]

In late April they reached Tonga, then known to the British as the Friendly Islands, where the *Bounty* stopped to take on supplies. Bligh's disciplinary outbursts had not abated. Christian approached the captain one day to offer a mild complaint: "Sir, your abuse is so bad that I cannot do my duty with any pleasure." As if to enact literally the farcical warning that "floggings will continue until morale improves," Bligh punished Christian for this remark. What Christian would call abuse continued and even increased as the crew made ready to leave the Friendly Islands. Some of the coconuts they had purchased ashore went missing, and Bligh raged violently and profanely at the crew, first Christian and then the rest of the men, calling them "rascals. . . . scoundrels . . . thieves," and "shrieking and waving his fists in the air, accusing them indiscriminately of theft." If this coconut incident pushed Bligh over the edge, it may have been his reaction to it that did the same for the crew. "There never was such a set of damned thieving rascals under any man's command before," Bligh raged. "God damn you, you scoundrels, you are all thieves alike . . . but I'll sweat you for it, you rascals. . . . I'll make half of you jump overboard before you get through the Endeavour Straits. You may all go to hell!" He then proceeded to order John Samuel, his clerk, to reduce the sailors' rations of food by half and to cut off their rum entirely. As it would turn out, the *Bounty* would never see the Endeavour Straits, a passage north of Australia and part of the ship's intended route homeward. Bligh was soon to be relieved both of his command and of the *Bounty* itself. "Flesh and blood cannot bear this treatment any longer," Christian was heard to say, just days before he would become the leader of a resistance.[14]

Before sunrise on April 28, 1789, the crew mutinied. As Bligh would later describe it in his defense of the circumstances by which he was to lose command of the *Bounty*, "Mr. Christian, with the master at arms, gunner's mate, and Thomas Burket, seaman, came into my cabin while I was asleep, and seizing me, tied my hands with a cord behind my back, and threatened me with instant death, if I spoke or made the least noise." The men bore cutlasses, muskets, and bayonets. They had secured the captain's cabin and confined any crewmembers thought to be loyal to Bligh to their own quarters. Bligh continued, "I was hauled out of bed, and forced on deck in my shirt, suffering great pain from the tightness with which they had tied my hands. I demanded the reason of such violence, but received no other answer than threats of instant death, if I did not hold my tongue." A launch, a small, deck-mounted boat, was prepared and lowered into the water from the ship.[15]

Quickly, frantically it seems from Hough's telling, the entire *Bounty* crew sorted themselves into two groups: those who would stay aboard and those who, for their loyalty to the deposed captain, would soon be set adrift. Arguments arose over which gear and provisions would be transferred to the launch and which would remain with the ship. Bligh himself was among the last to leave the *Bounty*. He tried, one last time, to avoid his fate. "Consider what you are about, Mr. Christian," he pleaded. "For God's sake drop it. I'll give my bond never to think of it again if you'll desist." The loyalists' and Bligh's final pleas began to register. The memories of wives and children were recalled. Christian was resolute.[16]

It soon became apparent, however, that the mutineers had miscalculated the popularity of their cause. Bligh's boat was overloaded with loyalists, and both Bligh and Christian realized that some men would have to be transferred back to the *Bounty* if anyone in the launch was to have even a chance of survival. Christian chose several who would stay with the mutineers. After some shuffling of positions, it was decided that eighteen sailors would join Bligh in the launch and the rest were to stay aboard the *Bounty*. To the loyalists for whom there was not room in the launch, Bligh offered reassurance: "Some of you must stay in the ship. Never fear, my lads, I'll do you justice if ever I reach England."[17]

The line was cut, and the launch began drifting away. Christian and his conspirators then proceeded to throw overboard every single potted breadfruit plant for which they had worked for so long under the Tahitian sun. There were more than a thousand. If Robert Dodd's painting of the event is accurate, the plants bobbed on the surface for a while before sinking, well within sight of Bligh, signifying the futility of his position.

Was this act merely a lightening of the load or did it carry deeper significance? Historian Jennifer Newell has hypothesized that "the tensions that had led to [the mutiny] had been exacerbated by the breadfruit. The ranks of green leafy stems had meant something quite different for the seamen" than they had for Bligh, and especially for Banks, comfortably ensconced back at Kew where he would remain ignorant of the mutiny for many months. Historians of the mutiny have described how the *Bounty*'s living quarters had been refitted to accommodate the breadfruit, forcing the crew to fit into tighter spaces than they had during the outbound voyage, and how the plants' need for irrigation had threatened to deprive the sailors of their freshwater rations. But most unforgiveable of all the breadfruit's offenses, according to Newell, was that, in being destined for delivery in the Caribbean, "the plants had taken them away from Tahiti." Now that the mutineers had wrested control of the ship, both the cargo and the royal orders were jettisoned. No information was shared directly with Bligh or those loyal to him regarding the *Bounty*'s new destination, but as the ship and the launch separated, Bligh heard the cry, "Huzzah for Tahiti!"[18]

Improbably, Bligh would in fact reach England. With strict rations, minimal navigational instruments, and cutlasses as their only weapons—"I asked for arms, but they laughed at me," Bligh would write in his *Account of a Mutiny*—Bligh steered first back toward one of the Friendly Islands.

Figure 6.1 The Mutineers turning Lieut Bligh and part of the Officers and Crew adrift from His Majesty's Ship the Bounty (29 April 1789), 1790, Robert Dodd, hand-colored aquatint engraving, 46 × 61.5 cm. (© National Maritime Museum, Greenwich, London.)

Finding there anything but a friendly reception, the deposed captain and his loyalists instead set a course for Kupang, a Dutch settlement on the island of Timor. After a month and a half of harrowing travel, in Bligh's words, "through the assistance of Divine Providence, we surmounted the difficulties and distresses of a most perilous voyage, and arrived safe in an hospitable port." From Kupang, Bligh traveled onward to Batavia, then to England, where a court-martial would completely exonerate him for the loss of the *Bounty*. Joseph Banks immediately began outfitting a second vessel, HMS *Providence*, to complete the breadfruit mission. This time the ship would be heavily armed and accompanied by another, the *Assistant*, so that any incipient mutinous talk could be promptly quashed. Bligh was given command of the *Providence* and a chance to redeem himself. This time he met success, and his "floating forest" arrived triumphantly at the harbor of Kingstown, St. Vincent, in early 1793—less than four years after the mutiny on the *Bounty*, fulfilling the dreams of Joseph Banks and a host of West Indian planters to have breadfruit, finally, in the Caribbean.[19]

In London in 2022, I boarded a red double-decker bus and rode out to the Lambeth neighborhood, immediately across the Thames from Westminster. My destination was the Garden Museum, set on the grounds of a deconsecrated church, St. Mary-at-Lambeth, the original structure of which was built nearly a thousand years ago. When Bligh died in 1817 he was buried in a family plot at the church, a block from his home. Today the museum's lively café features floor-to-ceiling glass doors offering an inviting view into the former church's courtyard. A tall tomb stands in the center. I read the mossy inscription—"we like them mossy," the museum's director has written—carved into one side of the structure: "Sacred to the memory of William Bligh Esquire, F. R. S. Vice Admiral of the Blue, the celebrated navigator who first transplanted the bread fruit tree from Otaheite to the West Indies, bravely fought the battles of his country, and died beloved, respected, and lamented, on the 7th day of December, 1817, aged 64." There was no mention of the mutiny, though one might read that account into the word "battles." The courtyard was full of all kinds of plants—it is, after all, the *Garden* Museum—but the tallest one, shading the tomb, was a fatsia, its big, lobed leaves so reminiscent of the plant that had infused Bligh's life with equal measures of suffering and glory that I first thought it was a breadfruit.[20]

After returning to Tahiti to reunite with their companions that had been left behind, the mutineers, under the command of Fletcher Christian,

Figure 6.2 Captain Bligh's tomb at the Garden Museum (formerly the church of St Mary-at-Lambeth) in London. (Photograph by Russell Fielding.)

eventually settled on Pitcairn, a remote Pacific island that had been charted inaccurately and would therefore, it was hoped, be hard to find for those seeking to bring them to justice. It was on Pitcairn that they decided to make a permanent home, as unmistakably symbolized—and realized—by their burning of the *Bounty*. A song by Canadian folk-rock band Captain Tractor titled "Pitcairn Island" tells the story of the mutiny on the *Bounty* from the perspective of the mutineers:

> Oh, we're sick of planting breadfruit,
> For the profit of our King.
> If we never see Great Britain again,
> We sure won't miss a thing.

To symbolize their commitment to island life on Pitcairn, at the moment the burning *Bounty*'s "bowsprit hit the waves," the song's melody changes from a classic sea shanty to the unmistakable backbeat of reggae. To this island-music rhythm, the last line of the chorus rings out one final time:

We'll dance the night away,
Underneath these foreign skies,
And we've never been this happy,
In all of our lives.[21]

Uninhabited at the time the mutineers made it their home, the island showed signs of previous human settlement, including the ubiquitous breadfruit trees, which grew all over Pitcairn. The descendants of the mutineers and their Tahitian wives and consorts comprise Pitcairn's population today. What is apparent from Pitcairn's history is that Captain Tractor took poetic license. Life on Pitcairn, post–*Bounty* burning, was frequently anything but happy. Alcoholism was common from the early days of the settlement, the mutineers having learned to distill a potent liquor from the *tī* plant, another of the Pacific Islanders' canoe crops that had been planted on Pitcairn. For such a small society, murders were remarkably frequent. Others died in accidents, of disease, or by suicide. When the settlement's existence was finally discovered and reported to British authorities in 1814, only one original mutineer, John Adams, survived. The Admiralty decided, mercifully, to grant Adams amnesty for his role in the mutiny. In 1823, thirty-three years after settling on Pitcairn, the community of descendants had deteriorated to the point that the crew of a visiting whaling ship took pity—and were maybe a little starstruck, considering how popular the story of the mutiny had become. Two crewmembers, John Buffett and John Evans, requested permission to leave the ship permanently, to settle on Pitcairn, and to help the islanders develop their society. Buffett was granted permission; Evans was not. Both men stayed anyway. They were the first new, permanent arrivals on Pitcairn in a generation.[22]

Despite the challenges, births outpaced deaths, and the Pitcairn population grew. Attempts were made to emigrate to Tahiti and Norfolk Island, but each time a remnant of the population returned home. After some of the would-be emigrants returned from Norfolk in 1859, life on Pitcairn began to settle down. Missionaries brought the Seventh-Day Adventist faith, which was adopted by nearly all the islanders. Peace and a gradual lessening of the island's isolation were the hallmarks of the twentieth century on Pitcairn. During the early twenty-first century, though, Pitcairn

Island was the focus of an international investigation and trial into the rampant and systemic sexual abuse of children, which the accused islanders claimed was simply part of their unique culture. As citizens of a British Overseas Territory, however, Pitcairn Islanders are subject to British law, and after a lengthy trial infused with history lessons from both sides, six Pitcairn men were found guilty of various charges related to their illicit sexual behavior with young Pitcairn girls and jailed.[23]

One might be inclined to think that modern Pitcairn Islanders would detest breadfruit, viewing it as the original cause of all their ordeals. Had it not been for breadfruit, there never would have been a *Bounty* voyage or a mutiny, and all the suffering on Pitcairn could have been avoided. As throughout most of the Pacific, though, breadfruit has become a dietary staple on Pitcairn as well. The fruit is harvested, however, in a way symbolic of the tenuous place it holds in the island's history. Unique among all the breadfruit-pickers of the world, Pitcairn Islanders today "hunt" their quarry with shotguns, blasting whole branches from the trees and retrieving the fallen fruits as though they were ducks in a marsh.[24]

CHAPTER VII

An Idea Whose Time Had Come

The novelist Victor Hugo is often credited with introducing the concept of "an idea whose time has come." While it may be the case that another French writer, Gustave Aimard, is more deserving of the credit, the phrase poignantly captures the emergence of a seemingly irresistible next step in human history. One such idea seems to have been the introduction of breadfruit to the Caribbean.[1]

The eighteenth century was a time in which the ideas of travel, exploration, and discovery seemed to be on every English reader's mind. Narratives of distant voyages were widely circulated and read in Britain and throughout the colonies. Literary historian Paul Fussell has written that the travel book was "one of the primary genres" of the time and that "the eighteenth-century literature we know would hardly be recognizable if we subtracted from it all its prevailing images of a rational and sturdy observer wandering about foreign parts, collecting data, patronizing the natives . . . and reporting his findings for the benefit of stay-at-homes."[2]

Some British Caribbean planters—among Fussell's "stay-at-homes,"—would have read the accounts of their contemporary eighteenth-century travelers by lamplight on the verandas of their great houses while frogs rhythmically chirped in the sugarcane fields. Reading would have served as a diversion from the pressures of plantation life, which, for the planter himself would have been of a different magnitude, trivial in comparison to the misery of the laborers he had enslaved, whose suffering was both existential and visceral. The planter's concerns, by contrast, were cerebral and economic: the price of sugar, the price of human beings, the debates in a Parliament an ocean away, and the day-to-day managerial decisions required to run a monocrop farming operation reliant upon forced labor.

One of these quotidian concerns was the feeding of the enslaved workforce. As island plantations expanded and sugarcane covered nearly all the land with agricultural potential, planters began to rely increasingly upon imported food for both themselves and their laborers. As more and more profits were generated, sugar plantations continued to grow to the point that, according to one late seventeenth-century government official, there was "not a foot of land in Barbados that is not employed even to the very seaside." The agricultural development on other Caribbean islands proceeded similarly. Every increase in the acreage under sugar required a corresponding decrease in the area of land available for other purposes, including the growing of provisions.[3]

Planters looked to the mid-Atlantic coastline of the North American mainland and began to import grains in such abundance that Delaware, New York, New Jersey, and Pennsylvania came to be known collectively as the "bread colonies." Soon this reliance became unsustainable. As early as 1708, Samuel Vetch, governor of Nova Scotia, wrote that no English island in the Caribbean was "capable of subsisting without the assistance of the [North American] Continent, for to them we transport their bread, drink and all the necessaryes of humane life . . . in so much that their being, much more their well being, depends almost entirely upon the Continent."[4]

But these imports were rich foods for rich planters. Enslaved and indentured laborers were fed cheaper meals: "dull and monotonous, consisting mostly of potatoes and boiled cornmeal," as historian Matthew Mulcahy has described the menu. Planters later recognized the need for something still cheap and plentiful, but also rich in protein and salt to fuel sixteen-hour days spent laboring in the cane fields and at the mills. Author Mark Kurlansky explained how cod from New England and Atlantic Canada was soon identified as the solution.[5]

"Badly split fish, the wrong weather conditions during drying, too much salt, too little salt, bad handling—a long list of factors" would result in a product unacceptable in the discerning markets of Europe, according to Kurlansky. Fortunately for the cod merchants, there was an entire region full of secondary consumers. "The West Indies presented a growing market for the rejects, for anything that was cheap," Kurlansky continued. Fish that might otherwise have gone to waste were eagerly bought by Caribbean planters to feed their enslaved laborers. A system arose by which ships would leave ports in the bread colonies with their eponymous resource, or harbors in Canada and New England laden with salt-dried cod, which came to be called "saltfish" in the islands, and return with cotton, indigo, salt, tobacco, and—most valuable of all—sugar from the Caribbean.[6]

Throughout the eighteenth century, Caribbean planters' reliance upon North American commodities deepened. By outsourcing their food production, planters were able to dedicate increasingly more land to their cash crops, growing richer by the acre. But the system would soon begin to break down. In response to the emergence of an independence movement in the soon-to-be United States, the British government imposed restrictive trade controls, severely limiting commerce between the North American mainland and its Caribbean island colonies. These restrictions drove a wedge down the middle of British America, which had been previously conceived as a single, integrated geographical region. Under England's new trade controls, the island colonies were suddenly forced into a greater degree of self-reliance. This was intended "to restrain colonial self-government," but had the unintended effect of creating food shortages—or at least the perception of food shortages—among the plantations of the Caribbean. Planters rarely considered that the native flora and fauna of these islands had been feeding their human inhabitants for generations before European arrival; probably they were very little acquainted with these native ecosystems after having largely replaced them with monocrop sugar fields. Rather, they appealed to the government, with one planter on St. Vincent writing that, "if the importation [of food] is not allowed we must inevitably Perish or bring in Provision without Permission." Leaders of the American independence movement recognized what power they had, owing to this dependence.[7]

Alexander Hamilton, the American statesman born on Nevis and raised in the then-Danish Virgin Islands, had worked during his youth as a clerk for Beekman & Cruger, a St. Croix-based mercantile firm. Hamilton's son John would write of his father that this experience had been "the most useful part of his education," as it developed his abilities in writing, accounting, and keeping inventories, as well as his knowledge of economics and geography—particularly of shipping routes. While these specific skills would serve Hamilton well as secretary of the treasury for the newly independent United States, the overall experience in maritime commerce impressed upon him how dependent the sugar islands were on the regular flow of goods from the north.[8]

Perhaps reflecting upon these formative years, Hamilton recognized how much leverage the soon-to-be independent American states held over the British Caribbean, and, by extension, Britain itself. British trade controls were met with American export embargos. Anticipating the repercussions of the Continental Congress's decision that, as of September 1775, "the exportation of all merchandize and every commodity whatsoever to Great Britain, Ireland, and the West Indies ought to cease," Hamilton wrote that "the West-Indians might have the satisfaction of starving." He contextualized, correctly, that "the lands in the West-Indies are extremely valuable because they produce the Sugar Cane, which is a very lucrative plant; but they are small in quantity and therefore their proprietors appropriate only small portions to the purpose of raising food."[9]

The prospect of starving was anything but satisfactory to the West Indian planters, and they acted quickly to avoid this outcome. As American independence and a shipping embargo loomed, some planters may have thought back to the travel books they had read on their verandas. The accounts of voyages to the Pacific had presented a view of a wholly different way of island life than the system of slavery and cash-crop agriculture in the Caribbean. Environmental historian Jennifer Newell has identified one book, John Hawkesworth's 1773 *Account of the Voyages Undertaken for Making Discoveries in the Southern Hemisphere*, as one of the more popular tales of exploration at the time.[10]

Hawkesworth himself was no explorer, but he compiled and edited the journals of four famous British explorers into one bestselling book. Included in his compilation was Cook's *Endeavour* voyage, which, owing to the accompaniment of Joseph Banks, produced some of the most glowing descriptions of breadfruit yet read. It is likely that some Caribbean planters would have read Hawkesworth, perhaps other contemporaneously popular accounts as well, and from these readings first developed a desire to have breadfruit growing in the Caribbean.

Hawkesworth described breadfruit as being "about the size and shape of a child's head, and the surface is reticulated not much unlike a truffle: it is covered with a thick skin, and has a core about as big as the handle of a small knife: the eatable part lies between the skin and the core; it is as white as snow, and somewhat of the consistence of new bread . . . its taste is insipid, with a slight sweetness somewhat resembling that of the crumb of wheaten-bread mixed with a Jerusalem artichoke." In the 1773 edition of Hawkesworth's *Account of the Voyages*, this description is interrupted by a page break, within which are two plates: one showing a view of Matavai Bay in Tahiti, and the other, a botanical illustration of a breadfruit branch with leaves, flowers and three large fruits.[11]

Later in the same volume, Hawkesworth paraphrased Banks, describing the livelihoods of the Tahitian people. "Of the many vegetables serving them for food," he wrote, "the principal is the bread-fruit, to procure which costs them no trouble or labour but climbing a tree. The tree which produces it, does not indeed shoot up spontaneously; but if a man plants ten of them in his lifetime, which he may do in about an hour, he will as completely fulfil his duty to his own and future generations, as the native of our less temperate climate can do by ploughing in the cold of winter, and reaping in the summer's heat, as often as these seasons return."[12]

These and other contemporary descriptions contributed to breadfruit's reputation among Caribbean planters who would have been interested in just such a niche commodity: not a rich, flavorful fruit but a bland, reliable, abundant and nearly labor-free source of energy for the workers they had enslaved. The planters were seeking not a delicacy for their own table but a fuel for their workforce. Historians Emma Spary and Paul White referred to breadfruit in the context of its eighteenth-century reputation among Europeans as "a superior staple." Literary scholar Elizabeth DeLoughrey has written that "planters insisted that the breadfruit would be a vital complement to the slave diet and had no intention of eating it themselves," citing one Jamaican planter who hoped it would serve as a "wholesome and pleasant food to our negroes." The planters' desire for breadfruit was economic—not gastronomic—and was encouraged by reading the reports of breadfruit's productivity, not of its flavor. Some may have concluded that breadfruit, if brought to the Caribbean, could relieve their sense of food insecurity and make their plantations more like those idyllic Pacific islands of their reading and their dreams. Of course, this was asking too much of a single plant. Historian Greg Dening laid out what he called "the basic paradox," that the breadfruit tree, "the very symbol of a free and unencumbered life, from the island of freedom, Tahiti," would be brought "to the islands of bondage, the West Indies and their slave plantations."[13]

When success eventually came, it was at great cost: two major, round-the-world voyages, one of which ended in the disaster of mutiny, spanning more than five years from the *Bounty*'s departure from England to the arrival of the *Providence* at St. Vincent. A modern Jamaican author, Michael Morrissey, has advised that "when we in the Caribbean are next enjoying the scent of roasting breadfruit, we must remember the many who lost lives and their loved ones to the breadfruit project of the late 18th century."[14]

Why were such imperial effort, expense, and loss given over to the transplantation of breadfruit from the Pacific to the Caribbean? The enslaved laborers on the British sugar islands already had a variety of starches including plantains and several imported root crops to serve as provisions. Even if another crop introduction was deemed necessary, why breadfruit and not something easier to obtain, more familiar, and from a closer source? Travelers' tales described so many plants of Africa, Asia, and Oceania that were new to Europeans, many of which were eventually brought to the Caribbean, but with far less public appeal or government-backed effort. Historian David Mackay referred to the eighteenth-century British clamor for breadfruit as "a collective national madness."[15]

Historians have long sought to identify who first conceived of this "madness"—the person who first had the idea to bring breadfruit from the Pacific to the Caribbean. Many have focused on Joseph Banks. There is no doubt that in the narrative of breadfruit's journey to the Caribbean, this gentleman-scientist who sailed with Captain Cook and would later serve as both director of Kew Gardens and president of the Royal Society was a central character. Mackay stated that "it may well have been Banks who suggested to the West Indians the idea of transplanting breadfruit." Similarly, historian Rebecca Earle wrote that "the conviction that the West Indies needed breadfruit seems to have originated with Joseph Banks."[16]

Banks first encountered breadfruit in Tahiti with Cook on the voyage of HMS *Endeavour*. One of his biographers, Toby Musgrave, wrote— probably with pun intended—that "the *Endeavour* voyage sowed the seed in [Banks's] mind of the potential economic rewards to be reaped from the acquisition and transfer of commodity plants to existing British colonies." While Banks's role in overseeing the breadfruit transfer project is well established, the historians who have attributed to him the origin of the idea are yet to bring forward documentation of his priority. Considering this absence of evidence, historian Jordan Goodman summarized his role well: "Banks, as one of the few people in England who knew anything about this plant, and who had actually been to Tahiti, was the perfect person to advise on the project, even though he did not initiate it."[17]

If the project, then, wasn't initiated by Banks, perhaps we can credit the person who first mentioned it to Banks in writing. That seems to have been Valentine Morris, a planter originally from Antigua and a personal friend of Banks. The author Charles Bucke called Morris "one of Nature's worthiest sons" and geographer Richard Howard noted that he was "an ardent horticulturist." Later, when Morris was appointed governor of St. Vincent, according to historian William Coxe, he "distinguished himself with . . . zeal and activity in promoting the cultivation of that island." It should come as no surprise, then, that Morris would appeal to Banks for aid with the horticultural improvement of Britain's Caribbean holdings.[18]

In April 1772 Morris was in London, and he stopped by Banks's Picadilly house for an unannounced visit. Not finding Banks at home, Morris left a note, which asked "whether there was no possibility of procuring the bread tree . . . so as to introduce that most valuable tree into our American Islands." If breadfruit could be brought to the Caribbean, Morris continued, he was "certain it would be the greatest blessing to the inhabitants." If his concern was for the *enslaved* inhabitants of Britain's "American Islands," however, it is worth noting that Morris, a slaveholder, might have instead considered emancipation as an even greater blessing. Morris's biographer Ivor Waters wrote that he had become intrigued with breadfruit after reading Banks's descriptions from his *Endeavour* voyage, which were widely available as compiled by Hawkesworth.[19]

Scholars have identified Morris's 1772 note to Banks, now archived at the British Library, as the earliest written suggestion of bringing breadfruit to the Caribbean. Historian Julia Bruce reasoned that because prior to Morris's 1772 letter, prominent publications promoting plant introductions had not included breadfruit among the "dozens of plants from all over the world [suggested] for transportation specifically to the American colonies," such an idea had not yet been put to print and concluded that Morris's letter is "the first known, written suggestion that breadfruit be transported to the West Indies." This assessment seems correct, when restricted to literature published in English and the correspondence conveyed within the British Empire, but it might not hold true within a broadened view of all relevant writing of the time.[20]

Overlooked in most historical breadfruit scholarship is the work of historian and philosopher François-Marie Arouet, better known by his *nom de plume*, Voltaire. Literary scholar Juliane Braun is among the very few to cite Voltaire in a study of breadfruit, remarking that, "in France, Voltaire idealized the plant as one that could 'serve to nourish and satisfy the hunger of humankind.'" Not included in Braun's citation was Voltaire's qualification that the potential nourishment and satisfaction offered by breadfruit would be realized only upon the condition that it "could multiply in other climates"—a clear suggestion of geographical transplantation, though without an explicitly defined destination.[21]

These remarks appear in a chapter titled "Arbre à pain" (tree of bread), in the first volume of Voltaire's 1770 *Questions sur l'Encyclopédie*—an obscure work that literary scholar James Hanrahan has called, "a significant Enlightenment text," but one that has "fallen between the cracks of editorial history." Scholars at Oxford University's Voltaire Foundation for Enlightenment Studies have noted that *Questions* is "Voltaire's longest work, and yet it is one of his least known." Looking into this obscure work, however, we see evidence of Voltaire's fascination with breadfruit and his ingenious, though veiled, suggestion that it might be transplanted to the Caribbean.[22]

Voltaire had not seen breadfruit himself but was familiar with the tree and its fruit through his own reading of the accounts of Pacific exploration, particularly the voyages of Anson and Dampier; both are mentioned by name in *Questions*. In the context of breadfruit's potential transplantation, Voltaire did not discuss the Caribbean region directly, but instead left readers a clue about the "other climates" he may have had in mind.

That clue was a reference to another botanical journey, one with which his contemporary readers would likely have been familiar. Before expounding further upon breadfruit, Voltaire raised the condition that its potential benefits would only be realized if it were to be "transplanted as the coffee tree was." How, and to where, was the coffee tree transplanted? Voltaire and his readers would have known the story that would later be called "the most romantic chapter in the propagation of the coffee plant"—the adventurous tale of its introduction to the Caribbean island of Martinique in 1723 by the French naval officer Gabriel de Clieu. When Voltaire suggested transplanting breadfruit "as the coffee tree was," he was likely referring to both de Clieu's adventure and his eventual Caribbean destination.[23]

Foreshadowing arguments later to be made by sugar planters, Voltaire predicted that breadfruit "could largely take the place of the invention of Triptolemus." In classical mythology, Triptolemus was the mortal hero who, having nursed at the breast of Demeter, the goddess of the harvest, taught agriculture to the Greeks. After suggesting that breadfruit might be transplanted, hinting that the Caribbean might be its destination, and claiming that its allegedly labor-free productivity could replace agriculture, Voltaire devoted the rest of the "Arbre à pain" chapter in *Questions* to a discussion of other staple crops of the world: wheat, maize, cassava, and rice. The discussion of these crops was intended to support Voltaire's assertion to his European readers that, although bread may be the "food to which we are accustomed" and may even be considered "a part of our being," it "is not the food of most of the world." A virtual tour then commenced, with Voltaire listing large and densely populated world regions—"all of southern Africa ... the immense Indian archipelago ... part of China"—which either "ignore bread" or in which "wheat is absolutely unknown." Placed as such within a brief essay extolling the benefits of breadfruit and suggesting its transplantation "as the coffee tree was," this denunciation of the necessity of bread may have led an attentive, late eighteenth-century Caribbean reader to ask: Why are we still importing expensive provisions from the bread colonies to fuel an enslaved workforce taken from Africa where, according to Voltaire, bread is ignored anyway? And what about this *arbre à pain*? Can there really be a tree of bread? And how might it be brought here "as the coffee tree was?"[24]

Voltaire's suggestion that the breadfruit tree be transplanted to the Caribbean, published in the first volume of *Questions* in 1770, predated Morris's letter to Banks by nearly two years. Thus, in terms of historical priority, we should recognize Voltaire, not Morris, as the first to express the idea in writing. But what about continuity? Is there any way this idea might have proceeded from Voltaire to Morris to Banks, culminating in the actual journey of breadfruit from the Pacific to the Caribbean? Certainly Morris might have come up with it independently by reading published accounts of Pacific voyages or through personal conversations with Banks; it does indeed seem to have been an idea whose time had come. Alternatively, the possibility remains that Morris may have been inspired directly by Voltaire.

Morris was educated; he studied at Cambridge, could read and write well in French, and was practiced in translation from French to English.

His friend and neighbor Philip Thicknesse considered Voltaire "his favourite author," referring to him as "the most agreeable writer, and the most intelligent Historian of this, or perhaps of any age." This was hardly a unique opinion at the time. The preface to a large collection of Voltaire's works, which Thicknesse owned, noted that the author's fame had spread "over all the civilized kingdoms and states of Europe" and remarked that, in particular, Voltaire "seems to be peculiarly adapted by nature, for the entertainment of the English people," citing as evidence the "eagerness with which his works are procured, translated, and perused by the natives of Great Britain." Thicknesse praised Morris as having "a good library" and since, as the literary historian Ifan Kyrle Fletcher has written, Thicknesse "frequently . . . loaned his friends copies of books he had just bought," he may well have loaned Morris the very books in which Voltaire suggested the idea of bringing breadfruit to the Caribbean.[25]

In the absence of definitive evidence that Morris read Voltaire and was influenced thereby to petition Banks to initiate the voyages that would bring breadfruit to the Caribbean, we can only summarize the clues suggesting that such a sequence may have occurred. Voltaire wrote in 1770 that, if it were be transplanted to the Caribbean, breadfruit "could largely take the place of" agriculture and would "serve to nourish and satisfy the hunger of humankind." Morris may have read these words on his own or his generous, book-loaning friend, Philip Thicknesse, who considered Voltaire "his favourite author," may have prompted him to do so. His familiarity with the Caribbean, along with his deep interests in horticulture and improving the lives of the poor, though blinded as he was to seeing beyond the cruel system of slavery, suggest that Morris was just the kind of person to take note of Voltaire's ideas on breadfruit. As a close friend of Joseph Banks, he was in exactly the position to petition for their implementation. On that April night in London, as he stood outside Banks's vacant home scrawling the quick note that would one day be preserved in his country's national library, Morris's thoughts may have been with Voltaire's *arbre à pain* and the new possibilities that this "most valuable tree" might find in "our American Islands."[26]

CHAPTER VIII

No Poem About Breadfruit

From the time of its introduction, breadfruit was treated with suspicion in the Caribbean. Horticulturally it did remarkably well in its new environment, but its acceptance into local cultures and cuisines was beset from the beginning by the way in which it was foisted upon an enslaved population. Three years after planting the first saplings in his island's rich, volcanic soil, Alexander Anderson, superintendent of the St. Vincent Botanical Gardens, sent an update to Joseph Banks regarding the trees' progress. "The Bread fruit thrives (if possible) better than in its native soil, all the trees in the Garden are full of fruit in all stages; they have been bearing constantly for 18 months past," he wrote. As for the fruits, "they are exceeding good boiled or sliced & toasted the same as Bread, and they make the finest pudding I ever tasted," Anderson wrote. His praise, however, came with a caveat: "strange to tell, there are some people who under value such a valuable acquisition, & say they prefer a plantain or Yam."[1]

Other observers and scholars have seen a deeper resentment toward breadfruit, particularly among those it was intended to feed. For example, historian John Parry sought to separate breadfruit's lukewarm reception in the Caribbean from what he perceived as the heroic efforts of those involved with its introduction: "It was not Bligh's fault that the Jamaican slaves at first refused to eat a plant which bore no resemblance to anything they had known in Africa or Jamaica. Breadfruit was fed to pigs for fifty years."

Similarly, art historian Jill Casid called breadfruit "the most infamous of colonial plant grafts" and wrote that it "became an edible plant so hated by slaves that by the mid-nineteenth century it was used exclusively for animal fodder." Historical geographer David Watts agreed, writing in his epic survey of the West Indies, that "for 50 years workers refused to eat it, preferring instead the more familiar Afro-indian combination of maize, yams, plantains and cassava, and leaving breadfruit to be fed to pigs and other animals."[2]

A food-focused reader might view these histories as signs of mere culinary preferences. People enslaved in the Caribbean already had plantains and other starches; when breadfruit was introduced, they preferred what they already knew. The first time I read Anderson's letter to Banks, my initial reaction focused superficially on the breadfruit and plantain merely as foods. Having grown up on the Cuban cuisine of southern Florida, I can hardly think of a better side dish than *plátanos maduros*, pan-fried to a perfect, sticky crisp. But remember the word of caution from historical geographer David Lowenthal: "in the Caribbean the past is a living presence." Because of its close association with the unmatched brutal injustices of slavery, breadfruit was—and in some cases still is—detested for more than just its bland taste.[3]

The fifty-year period referenced by historians is directly relevant to the history of slavery in the Caribbean. The passage of the Slavery Abolition Act of 1833, which took effect in 1834, converted every person enslaved on a British Caribbean island into an "apprentice." These required apprenticeships were intended to last until 1840, but owing to widespread opposition to their false sense of freedom, they were ended two years early, in 1838. Considering that breadfruit arrived in the Caribbean in 1793 and was distributed throughout the region over the following years, its fifty-year period of feeding pigs rather than people roughly corresponds to the remaining years of slavery and the euphemistic apprenticeships in the region.

Some in the Caribbean have argued that the period of breadfruit antipathy lasted much longer; some would have it extend into the present. Gordon Rohlehr, a scholar of Caribbean culture and music, added at least another century, writing with reference to a calypso song from the 1930s, that "breadfruit was at the time scorned as a low status food, fit mainly for pigs and 'dem small islands.'" Extending the anti-breadfruit period even further, into the 1990s, the poet Édouard Glissant wrote that many on his native island of Martinique hated breadfruit as children, owing to it being "intimately associated with the idea of poverty and the reality of destitution."[4]

Author Jamaica Kincaid wrote of breadfruit in 1999, "no West Indian that I know has ever liked it." This was not her first critical breadfruit remark. Kincaid's 1986 novel *Annie John* features a young Antiguan girl who resents her mother for tricking her into eating "the much hated breadfruit." *Annie John* doesn't include the backstory of why the titular character hated breadfruit. But Kincaid provided a clue regarding her own thoughts on the matter in *My Garden (Book)*, her nonfiction botanical journey through the gardens of the Caribbean and New England. Here she wrote, "This food, the breadfruit, has been the cause of more disagreement between parents and their children than anything I can think of. . . . It was meant to be cheap food for feeding slaves. . . . It grows readily, it bears fruit abundantly, it is impervious to drought. . . . In a place like Antigua the breadfruit is not a food, it is a weapon."[5]

Kincaid's metaphor rings literally true in the words of Hawaiian poet Wayne Kaumualii Westlake, who wrote,

breadfruit
what's that?
the tourist gasped—
should'a stuffed one
in her face!

Westlake had imagined using breadfruit—'ulu, a Hawaiian canoe crop—violently to express frustration with outside cultural influence in Hawaiʻi. Unfortunately, someone acted on his "should'a," in 2017, resulting in an injured tourist and widespread concern about increasing anti-tourism sentiment in Hawaiʻi. A tour bus was about to depart the Polynesian Cultural Center on Oʻahu when suddenly a window burst, and something struck a passenger in the face. The victim had been hit with a small, immature breadfruit thrown with enough force to break both the window glass and her nose. Not a food, a weapon.[6]

But breadfruit's more common use as a weapon has always been political. The colonial governments of the Caribbean attempted to use it to prop up their plantation system and the forced labor it required while political change and shifting food systems threatened their continuation. For as long as breadfruit was meant to support the scourge of slavery, it remained "much hated." Thus, the Slavery Abolition Act of 1833 marks the beginning of breadfruit's redemption and its first tentative step toward acceptance in the region. Historian Mathieu Morin stated plainly that "breadfruit never became the important staple it was meant to be during the slavery period," and that colonial efforts like the transplantation of breadfruit from the Pacific to the Caribbean failed to achieve their goals "because they could never solve the real problem, *which was slavery.*" Slavery was a deplorable crime, humanity was its victim, and, as Kincaid described it, breadfruit was one among an arsenal of weapons.[7]

After emancipation, the relationship with breadfruit got more complicated. On the one hand, the acceptance—and even the embrace—of breadfruit was part of a larger act of reclamation on the part of newly free Caribbean people. On the other, since breadfruit's mere presence in the Caribbean region was so clearly and directly tied to slavery, it retained a thick green stigma of scorn. Ninety-nine years after the end of the apprenticeships, the Caribbean Black nationalist Marcus Garvey would speak of the need "to emancipate ourselves from mental slavery because whilst others might free the body, none but ourselves can free the mind." To some in the Caribbean, the sight of breadfruit trees and sugarcane were part of the "mental slavery" from which they still sought emancipation. Decades later, Garvey's words would inspire one of Bob Marley's best and best-known tunes, "Redemption Song."[8]

The phrase may sound simple when accompanied by Marley's acoustic guitar, but mental emancipation and the true freeing of one's mind is a staggeringly complex endeavor, especially when played out on the national scale against so great an adversary as the combined forces of colonialism and slavery. Historian Hilary Beckles has written convincingly about "how Britain underdeveloped the Caribbean." In his 2023 book by that title, Beckles argued that Britain, through the colonization of the Caribbean and the reliance upon two hundred years of unpaid labor, extracted massive wealth from the region while simultaneously and intentionally creating impoverished societies dependent upon European aid and subverting any local initiatives by which the Caribbean people would have prepared themselves and their nations for emancipation and eventual political independence. Beckles's solution? The UK should pay.[9]

Whether or not the goal of economic reparations for slavery will ever be achieved in the Caribbean, or the rest of the Americas, is yet to be seen.

But from the moment of emancipation, those who had been enslaved began to claim ownership of the former apparatus of their oppression, beginning the process that Garvey later prescribed and about which Marley later sung. New cultures, languages, and political entities were created from the wreckage of an empire built upon slavery-dependent plantation agriculture. An act of Parliament had freed the body; none but those formerly enslaved would free their own minds. One necessary step toward mental emancipation was a collective exercise in reclaiming and reframing that saw former plantation houses turned into public buildings, former "slave talk" dialects recognized as legitimate creole and patois languages, and the culinary symbols of slavery—ground provisions, saltfish, and breadfruit, among others—admitted, on a sentimental as well as a practical basis, into the diets of free people.

The abolition of slavery was, in the words of historian Demetrius Eudell, an "incomplete victory." Those who had been enslaved became merely colonized. For more than a century, there was talk—both in the columns of British newspapers and under the shade of Caribbean breadfruit trees—about whether people were actually better off post-emancipation or if the Slavery Abolition Act had been nothing but a "false Canaan." Eudell employed this biblical imagery to consider whether the newly free people would have been better off having been allowed to "remain in Egypt in our bondage" than having been imperfectly freed. This line of questioning was, of course, rhetorical. The failings of emancipation were not evidence that slavery should have continued, but rather that freedom should have been done better.[10]

Later, as waves of independence washed over the region during the 1960s–1980s, the former British islands embarked on the journey of nation-building, a process that required a serious and renewed consideration of what to do with the leavings of empire. Like survivors fleeing a sinking ship, these now-independent Caribbean citizens made difficult choices, sometimes without unanimity, about what to keep and what to let go down in the whirlpool. In most cases, breadfruit was given space on the lifeboat where, just like it had done for generations of voyagers navigating other seas, it would provide sustenance to those courageous enough to set out into unknown waters and establish new island societies. But because of its past, breadfruit was treated with the suspicion of a mutinous crewmember.

This was true in the former French colonies as well. Biologists Gerald and Selmaree Oster, who had spent years working in rural Haiti, wrote in the 1980s that breadfruit's "stigma of the past" had led many Haitians to consider it "only fit for pigs and country bumpkins." The reference to bumpkins is especially apt, as breadfruit's post-emancipation stigma in the Caribbean rarely reached Kincaid's level of hatred but instead was typically expressed merely through ridicule and derision.[11]

In a 1978 interview the poet Derek Walcott explained this perspective. Breadfruit, he said, along with other fruits grown in the Caribbean, "used to have a comic association because you were taught to look at these things as inferior and inferior things are comic." Reflecting upon his early education, Walcott recalled "having been taught that there's a literature that talks about the cantaloupe, the melon, the apple, and the peach but that there's no poem about breadfruit." The result of this literary exclusion, he concluded, is that one "begins to feel that despite the fact that this particular fruit sculpturally is more interesting, that it carries an inferior association." That association led many in the Caribbean to deride breadfruit, to consider it—and by extension, themselves—inferior to the people and things hailing from more temperate climates, the literarily superior fruits like the apple and the peach.[12]

At least the derision of breadfruit's "inferior association" is the major way that *Black* Caribbean people responded to its stigma. White Caribbean people, for more than a century following emancipation, took a different view. To them, breadfruit's predominant flaw was not that it had abetted slavery but, echoing the views of early European explorers in the Pacific, that its abundant productivity and easy maintenance made people lazy. For this, rather than deriding breadfruit, they wanted to destroy it. Of course, this assumes a monochromatic Black-white dichotomy that greatly oversimplifies Caribbean perceptions and lived experiences of race and skin color. The reality is more complicated.[13]

In 1947 and 1948, a trio of European travelers spent six months island-hopping from Cuba to Trinidad, collecting stories and photographs for what would become *The Traveller's Tree*, a book described as "a report on the birth pangs of our postwar world just then coming to be." The mid-twentieth century was an auspicious time to travel in the Caribbean. Nearly a hundred years past emancipation but not yet into the era of postcolonial independence, most islands were still in the liminal state of being colonized possessions inhabited by free individuals. Slavery remained bold in the collective memory, and for most island nations political independence loomed promisingly on the horizon.[14]

Patrick Leigh Fermor, one of those travelers and the author of the account of their journey, remarked that "every single one of the inhabitants of these islands had grandparents or great-grandparents that were either owner or slave; often both; and the few generations that have elapsed since the emancipation have not yet been sufficient to blot out the effects of those three hundred abnormal years." The short phrase "often both" is what to me stands out the most in this line. Based upon modern, simplistically racialized thinking, we tend to suppress the diversity present within any one person's genetic and cultural background.[15]

The most poignant and personal account of the complexities of Caribbean ancestry that I know is the book *Sugar in the Blood*, by Andrea Stuart. Stuart illustrated the arc of Caribbean history, spanning from slavery to emancipation to decolonization, by telling the meticulously researched story of her own Barbadian family, which is, in her words, "mixed-race on both sides, blending the histories of both oppressor and oppressed." Every one of us, genetically speaking, is made up of the many who went before. This is just as true in the Caribbean as anywhere else, but there it may be more portentous than in most other world regions since a Caribbean person's ancestry might include, like Stuart's, the names of both enslavers and enslaved, those whose efforts brought the breadfruit to aid their own cruel system of self-enrichment, as well as those who, for the sake of their own dignity, refused to eat it.[16]

Perhaps breadfruit performed its duty too well, relieved too much labor in the provisioning grounds, because soon after emancipation, it began to receive blame for enabling people's—particularly Black people's—laziness. To some, it would seem, the abundance of a breadfruit tree's productivity degraded the spirit even as it nourished the body. James Mitchell, prime minister of St. Vincent and the Grenadines from 1984 to 2000, infamously criticized what he referred to as "the breadfruit mentality" of his compatriots, particularly those "who wished to reap without sowing." Mitchell explained in his autobiography that "the wonderful tree that Captain Bligh had suffered to bring us required absolutely no attention, no fertilizing, no pruning, no spraying. It is a tree to be reaped without any care." Trained as an agronomist at the Imperial College of Tropical Agriculture in Trinidad before getting into politics, Mitchell would have known he was being hyperbolic. Still, the "breadfruit mentality" accusation became Mitchell's curse—his equivalent to Nixon's "not a crook" or Clinton's "that woman"—and is still discussed in the Caribbean today as an example of political condescension toward the common citizen.[17]

The name *breadfruit* itself may lie subconsciously at the heart of the problem. This compound word uneasily joins an internal paradox that calls to mind the tilling, milling, kneading, proofing, and baking required for bread, and, simultaneously, the free-gift nature of fruit. Bread has long been a symbol of human industry, precisely because of the effort required to make it. Bread only became possible when our ancestors turned to agriculture, abandoned their nomadic life, and stayed in place long enough to sow and reap a season's domesticated produce. This settling naturally required social organization and the association between the system and the product was made clear by Benjamin Franklin, who wrote, "Where there's no law, there's no bread." In praying for God to "give us this day our daily bread," Christians are asking that this food be meted out, individualized, and owned.[18]

Fruit, by contrast, symbolizes all that may be hunted and gathered, foraged from the natural world without payment because there is no one to pay. Fruit, biblically speaking, symbolizes Eden before the Fall, an antediluvian epoch in which God freely gave people "every tree, in . . . which is the fruit of a tree yielding seed." This was the notion behind the assertions of Banks and Cook that the Tahitians were an uncursed people. Others have pushed back against the prominence of bread and the private capital it represents. The Enlightenment philosopher Jean-Jacques Rousseau, for example, in his warning to view critically all claims of private property, admonished his reader, "You are undone if you once forget that the fruits of the earth belong to us all, and the earth itself to nobody."[19]

Bread and fruit, then, might stand as comestible symbols for the two major clades of humanity identified by Daniel Quinn in his novel *Ishmael* as the Takers and the Leavers. Takers make up the cultures that have been through both the agricultural and industrial revolutions; Leavers are those few remaining societies who have not. Quinn's Takers make the rules, and, when it comes to foraging for fruits, the rules are largely prohibitive, treating the collection of wild food almost as if it were praedial larceny. And while scholars have shown the manifold negative effects on society, including "social and sexual inequality . . . disease and despotism," resulting from the agricultural revolution—which geographer Jared Diamond has called "the worst mistake in the history of the human race"—we maintain a preference for food production over food collection in both the structure of our mainstream societies and the themes of our origin stories. Adam and Eve were condemned for partaking of the forbidden fruit; Christ is the bread of life.[20]

What, then, can it mean for "bread" to grow on trees? While Pacific Islanders had known it for centuries and Europeans had previously described it, breadfruit got its English name—and its association with bread began—when William Dampier, the seventeenth-century English pirate, penned his description of "The Bread-fruit." Dampier stated matter-of-factly that Pacific Islanders "use it for bread . . . they bake it in an Oven . . . scrape off the outside black crust . . . and the inside is soft, tender and white like the crumb of a Penny Loaf." Fair criticism has been made of what literary scholar Vanessa Smith has called the "British rhetorical insistence on the fruit's bread-like qualities," an insistence first made in the name Dampier coined and perpetuated through centuries of comparisons between breadfruit and bread. As a point of pomological fact, breadfruit is fruit; it is not bread. Nor does it *strive* to be bread in the sense of some Platonic ideal and form. Breadfruit is not the bread of the Caribbean or Pacific Islanders. Just as Tolstoy is the Tolstoy of the Zulus and Proust is the Proust of the Papuans, *bread* is the bread of the Caribbean and the Pacific. Breadfruit is the breadfruit of the world, unless, to apply journalist Ralph Wiley's critique, "you find a profit in fencing off universal properties of mankind into exclusive tribal ownership."[21]

But throughout the period of European exploration in the Pacific, the idea that breadfruit was trying, but just failing, to be bread remained current as people continued to "fence off" breadfruit as the property of the Pacific Islanders and bread as the property of Europeans. The attorney James Boswell recorded a conversation between himself and author Samuel Johnson in 1773 in which he, Boswell, asserted that, "I am well assured that the people of Tahiti who have the bread-tree, the fruit of which serves them for bread, laughed heartily when they were informed of the tedious process necessary with us to have bread: plowing, sowing, harrowing, reaping, threshing, grinding, baking." To this, Johnson replied, condescendingly, "Why, Sir, all ignorant savages will laugh when they are told of advantages of civilized life. Were you to tell men who live without houses, how we pile brick upon brick and rafter upon rafter, and that after a house is raised to a certain height, a man tumbles off a scaffold and breaks his neck, he would laugh heartily at our folly in building; but it does not follow that men are better without houses. No, Sir, (holding up a slice of a good loaf,) this is better than the bread-tree."[22]

To view breadfruit as a bread substitute is to diminish both bread and breadfruit. It is to assume a socio-culinary hierarchy that does not exist. Six days a week at the Cheapside Market in Bridgetown, Barbados, vendors stack pyramids of locally harvested breadfruit for sale. Just down the street and around a corner, Crumbz Bakery serves up fresh loaves of salt bread, ready for slicing and making flying-fish cutters, a handheld version of the Barbadian national dish. No one would confuse the fruit of the market with the bread of the bakery. Each is perfect in its own role.

An exemplary case of breadfruit-as-bread thinking occurred on a Sunday in 1797, when it was reported that at an Anglican Eucharist held by missionaries deep in the South Pacific, "for the first time the bread-fruit of Tahiti was used as the symbol of the broken body of our Lord, and received in commemoration of his dying love." The missionaries had replaced the traditional communion wafer with slices of breadfruit. This report drew swift condemnation from the Missionary Society's London headquarters, which called the substitution a "desecration of the most sacred rite of Christianity" and wondered rhetorically if the missionaries had also "used palm wine in preference to the juice of the grape." During the very same decade when breadfruit was first brought to the Caribbean, not everyone was ready to accept breadfruit as bread.[23]

Bread and breadfruit may be as different, yet onomastically related, as Mark Twain's "lightning bug and the lightning," but the association endures, carrying its implications from the seventeenth century to today. Europeans and their postcolonial descendants have long held tropical peoples in contempt for their supposed "breadfruit mentality" and its attendant laziness, a trait assumed to be enabled by the bountiful botanical environment in which they live. A nineteenth-century American geography textbook, in a section titled "What Men Do–Occupations," began its description of the varied ways around the world that people work with a simple contrast: "We cannot live as some men do in hot countries, where bread grows on trees, and milk may be had from the cocoa-nut. We must work hard for what we get."[24]

Western, educated, industrialized, rich, and democratic societies—a set of descriptors humorously acronymized as WEIRD—are built on principles that value hard work. We want to think we've got it better: both when compared to others and to our own past. Surely the agricultural revolution was worth it. Are not those few remaining nonindustrial societies, still living by hunting and gathering, relying upon the free gifts of nature and easily cultivated crops like breadfruit, missing out on something? Maybe. The seventeenth-century political philosopher Thomas Hobbes characterized life "in the state of meer nature," famously, as "solitary, poore, nasty, brutish, and short."[25]

Contrasting this bleak view, though, is the idea, first explored by anthropologist Marshall Sahlins, that preagricultural peoples might have been "the original affluent society." Sahlins showed evidence for his assertion that the adoption of this philosophy and the lifestyle that follows can lead to shorter workdays and more time for leisure than typically experienced in industrial societies. Hunter-gatherers, according to Sahlins, had found a "Zen solution to scarcity," by which "a people can enjoy an unparalleled material plenty." Less work, more leisure time, more multigenerational family contact: Sahlins argued that taking one's food from nature can, in the right mindset and ecological setting, offer a path to "affluence without abundance." Life under the breadfruit tree may indeed be affluent. And that might be precisely why industrial societies have kept their distance.[26]

People living in WEIRD societies are taught to recoil at the thought of what breadfruit symbolizes: a free lunch. Our economists warn us explicitly that there is in fact no such thing. Is this why people in Western, developed countries have largely excluded breadfruit from their diets? Temperate countries import other tropical products—cocoa, coffee, palm oil, rubber, and sugar are just a few examples—but almost no breadfruit. Perhaps it is our knowledge of the intense labor that goes into the production of these commodities, contrasted with the largely labor-free reputation that breadfruit has acquired. While I have occasionally encountered frozen breadfruit slices and food products with breadfruit as an ingredient in the frozen food aisle, I have never seen fresh, whole breadfruit for sale at a grocery store or farmers market in the mainland United States.[27]

If the "breadfruit mentality," the wish "to reap without sowing," is indeed a real problem, would ridding the Caribbean islands of breadfruit trees be the solution, just as it was suggested by Coleridge regarding the Tahitians? The idea has been proposed. "I'd like to see them all cut down, every one of them, and burnt," a government official in Martinique told Fermor and his companions as they were researching *The Traveller's Tree*. "It's the bloody bread-fruit that keeps the black alive without working. It lets them grow fat without doing a hand's turn, takes away all their incentive to work. That's what puts them beyond our control." Fermor went on to describe how the official "held up two hands, with the fingers crooked like claws through which the whole of the Negro world was slipping." He then addressed Fermor directly, "And who's to blame for *that? You*, sir." At this point, Fermor narrated, "A claw was placed on one of my shoulders in a conciliatory gesture." The official continued, "Not you directly, but your Bligg!" "Bligg?" wondered Fermor. "It took him some time to make it clear to us that the real villain of this disaster was the captain of the *Bounty*, none other than Captain Bligh, who first brought the bread-fruit tree from the East Indies and planted it in the Caribbean." The Martiniquais official then, in a manner familiar to emotional bilinguals, angrily switched to his native French: "*Bligg est le coupable, messieurs, de tout ça. Ne me parlez pas de l'arbre à pain!*" Bligh is the culprit, sirs, for all of this. Don't talk to me about breadfruit!²⁸

The official's wish, that the breadfruit trees of Martinique might be "cut down, every one of them, and burnt," would never happen. It's common knowledge in the Caribbean that you don't cut down a breadfruit tree unless it's stopped producing or died. Why would you? What could possibly take its place that would even approach its bounty in terms of food, shade, beauty, and habitat? An article in a local St. Vincent and the Grenadines newspaper advised that "to cut down a man's breadfruit tree is to wipe out the very means of his existence." Indeed, the phrase is used metaphorically throughout the region to describe any action that keeps another person from making a living. If you sink a fisherman's boat or smash a seamstress's sewing machine, you have, in the colloquial parlance of the Caribbean, cut down their breadfruit trees.²⁹

Another reason to leave breadfruit trees standing is for the mute witness they bear to the Caribbean's past. A correlation may exist between the modern density of breadfruit trees on any given Caribbean island and the magnitude of that island's enslaved population during the plantation era. An early twentieth-century evaluation of the Vincentian economy, which prior to emancipation had been briefly but deeply dependent upon enslaved labor, found that "valuable foods grow as free gifts of nature" and offered as proof: "the ratio of breadfruit trees to [human] population is probably greater than in any other West Indian island." By contrast, the handy botanical guidebook, *Plantes utilitaires de Saint-Barthélemy* (Useful Plants of St. Barthélemy) lists the abundance status of "*Fruit à pain*" as "cultivated very rarely in gardens," and laments that, "the species has almost disappeared" from the island.[30]

St. Vincent and St. Barthélemy are quite different islands in many ways, particularly regarding their histories. While slavery was practiced on both islands to be sure, it is interesting to note that in 1830, the enslaved population of St. Vincent was 83 percent of the island's total but comprised only 35 percent of St. Barthélemy's population. More people enslaved by the cruel plantation system seems to have meant more breadfruit trees. Whether present-day breadfruit coverage might be correlated with nineteenth-century enslaved populations on other islands is yet to be proven. It may be the case that the breadfruit trees growing today in the Caribbean do so in silent, quantitative testimony of the degree to which slavery tainted each island's history.[31]

On those Caribbean islands where breadfruit trees are most prevalent, it may have been only a matter of time before their appeal—both utilitarian and aesthetic—began to win over the population. Eventually, inevitably it seems, the memory of breadfruit as "cheap slave food" and "a weapon" in service of colonialism and slavery would soften and its sustenance of both life and land would emerge. As with so many such subtle transformations, it would take a poet to put this gradual shift to words. History, it seems, preordained that the task would not be undertaken by just any poet, but by St. Lucia's Nobel laureate, Derek Walcott.

According to literary scholar Erika Smilowitz, Walcott's long game, the "conscious preoccupation" of his career, was "to verbally reclaim the Caribbean landscape" by "creating a non-colonial literary perspective." Literary scholar John Thieme would agree, noting Walcott's effort "to invest trees such as the breadfruit and mango with the same status as the elm and the oak and to reclaim ordinary St. Lucian lives from what Walcott has referred to as the legacy of the 'nameless, anonymous, hopeless condition' of slavery." Walcott achieved this, in part, by rejecting the "inferior association" that haunted breadfruit and, in Thieme's words, by suggesting that breadfruit, which he called "the legacy of the *Bounty*" might be "re-envisaged from a very different viewpoint." This new viewpoint is one in which "the understandably maligned breadfruit undergoes transubstantiation" to become something "no longer simply a metonym for a disparaged Caribbean reality, but a quasi-spiritual substance, which . . . carries the potential for creative transformation."[32]

Why was breadfruit given such grace, considering its role in the brutality of slavery? According to literary scholar Hannah Rachel Cole, it may have had something to do with the mutiny aboard the *Bounty*. Cole has referred to Walcott's "poetic account of the mutiny [which] reveals how the breadfruit trees themselves participated in the act of insubordination that delayed their importation to the British West Indies for four years." Just like the Africans who had been kidnapped from their homes, shackled, brutalized, and shipped across the Atlantic to face short lives of misery, breadfruit too had been taken from its home by colonial Europeans, transplanted and made to work, producing food without compensation, without participation in the *kuleana* that it knew in the Pacific. In this view, breadfruit might be seen less as co-conspirator with the enslavers and more as a co-sufferer with the enslaved in "a system of multispecies forced labor." Or, as Cole put it, "people and plants alike . . . located in the wake of colonial transplantation."[33]

Walcott's rejection of the associations born out of colonialism, racism, and slavery, as well as his direct reconsideration of the breadfruit, can be seen most explicitly in his 1997 epic poem *The Bounty*, which he dedicated as an elegy to his mother, who died in 1990. In this poem, named for the ill-fated ship, but also for the literal bounty—the abundance—of food that breadfruit represents, Walcott imagined a scene unfolding as the mutiny began in which,

> Faith grows mutinous. The ribbed body with its cargo
> stalls in its doldrums, the God-captain is cast adrift
> by a mutinous Christian, in the wake of the turning *Argo*
> plants bob in the ocean's furrows, their shoots dip and lift,

Walcott, in other words, imagined the breadfruit saplings as suffering alongside both the pained mutineers and the deposed captain and his loyalists. Suffering all around.[34]

In an earlier line Walcott had called breadfruit, "*bois-pain*, tree of bread, slave food." The first of these names, *bois-pain*, is a term for breadfruit from the French-based Creole language of St. Lucia; Cole saw it as "a bilingual pun," suggesting that breadfruit is a "tree of pain." Whether that pain is felt by people or by trees themselves is left up to the reader. The entire poem is one of pain, of melancholy, but the mood brightens when Walcott imagines his mother, along with all "the beloved [who] have vanished," being

> relieved of our customary sorrow,
> they are without hunger, without any appetite,
> but are part of earth's vegetal fury;

The poem then beautifully portrays the grieving son finding traces of his mother in "the wild mammy-apple, the open-handed breadfruit / . . . the open pomegranate . . . the sliced avocado." It is through these fruits—"a medley of imported and native species," according to Cole—their role elevated to that of the memory of a departed loved one, that Walcott imagined being still able to commune with his mother even after her death.[35]

In writing *The Bounty*, Walcott captured the elevation of breadfruit above its former status. Twenty-one years earlier, in his 1978 interview, he had referred to breadfruit's "inferior association" as compared to "the cantaloupe, the melon, the apple, and the peach" and lamented that while those temperate fruits had all been represented in the literature he was taught in school, "there's no poem about breadfruit." In mourning his beloved mother's death and eulogizing her memory, St. Lucia's Nobel laureate wrote the breadfruit poem the world was lacking.[36]

CHAPTER IX

Fruits of the Creole Kind

For its first six years as an independent nation, from 1979 until 1985, St. Vincent and the Grenadines was represented by a flag showing the country's coat of arms over a breadfruit leaf. That the leaf of a fruit-bearing tree originally imported from the other side of the world to feed their enslaved ancestors would be chosen as a unifying symbol for an emancipated and newly independent Caribbean country is evidence of a remarkable transformation. Breadfruit's shift from V. S. Naipaul's characterization as "cheap slave food" into a symbol of pride and independence is a story of creolization, of "becoming indigenous to a place," a transformation that, according to ecologist Robin Wall Kimmerer, results from "actions of reciprocity, the give and take with the land." In the Caribbean, the breadfruit tree—like the people who grow it and eat its fruit—has creolized and is on its way to becoming indigenous.[1]

Cuisines, cultures, languages, and people themselves can all be *creole*. The word's first recorded use was by William Dampier, the same pirate and adventurer who also gave breadfruit its English name. Dampier described "one Mr. Cook, an English Native of St. Christophers," as the island of St. Kitts is formally known, as "a Criole" and, recognizing his reader's likely unfamiliarity with the word, explained that it referred to "all born of European Parents in the West Indies." The *Oxford English Dictionary* notes that the term is used "chiefly in the Caribbean" to distinguish people born of European or African parentage "from those of similar descent who were born in Europe or Africa, and from indigenous peoples." In the plantation-era Caribbean, the word meant "born-here," specifically distinguishing creoles from those who "came-here" in the case of free people or were "brought-here" in the case of those who were enslaved. Under this definition, a family would creolize in a generation. And just as with babies, the birth of new cuisines, cultures, and languages could usher in the creolization of entire societies.[2]

One area in which creole cultures of the Caribbean have particularly flourished is in the development of new Afro-Caribbean religions. Some of the more familiar examples of these creolized religious traditions are Vodou in Haiti, Santería in Cuba, and Rastafari in Jamaica. The last of these, Rastafari, or the Rastafarian movement (never Rastafarian*ism*, owing to Rastafari's focus on unity and the negative association between "-ism and schism"), originated during the early twentieth century in Jamaica as an Afrocentric, particularly *Ethiopia*-centric, community of believers in the foretold coming of a Black Messiah. Many Rastafarians believe that this prophecy was fulfilled in 1930 with the coronation of Emperor Haile Selassie in Ethiopia, who, before attaining the rank of *Negusa Nagast*, King of Kings, was known as Ras Tafari, *ras* being a title similar to *prince*, and *Tafari* his given name. While the Rastafarian movement is not a Christian denomination, the Christian Bible—including both the Old and New Testaments—is venerated, read, and, when deemed necessary, reinterpreted as Rastafarian scripture.[3]

The influence of biblical teaching on the Rastafarian movement can be seen particularly in the concept of *livity*. This broadly encompassing term describes the Rastafarian way of life, which is held up, according to sociologist AlemSeghed Kebede, as "an alternative to the lifestyle of most members of society." *Livity* influences nearly all aspects of the lives of Rastafarians, starting, perhaps most basically, with food. Many Rastafarians adhere to a set of dietary customs that promotes the consumption of *ital* food. The terms *ital* and *livity* are both part of Rastafari's consciously constructed language, Iyaric. Based upon Jamaican Patois, Iyaric features—among other attributes—the replacement of individual syllables carrying negative connotations with alternative, positive ones. For example, the English word *diet* includes the syllable *die*. Since dying is almost always perceived as negative, Iyaric replaces the syllable with *live*, and thus the Rastafari word for one's nutritional regime—one's diet—becomes *livit*. Iyaric also features a preference for so-called I-words, frequently replacing syllables with "I" as a reference to the Iyaric phrase "I and I," defined by sociologist Ellis Cashmore as "an expression to totalize the concept of oneness," expressing unity among speaker, listener, nature, and God. A relevant example here is the replacement of the word *vital* with *ital*. *Ital* food, then, is food that is vital; a *livit* based upon *ital* food promotes vitality and is thus part of *livity*.[4]

The *ital livit* draws heavily upon Rastafarian interpretations of Hebrew dietary laws, including both standard Kashrut and the proscriptions associated with the Nazarite vow, the latter of which forbids both alcohol consumption and the cutting of hair, which is why Rastafarians often embrace the mind-altering effects of marijuana but usually avoid drinking alcohol and wear their hair in dreadlocks. *Ital* food is primarily vegetarian, often vegan, and is believed to promote both physical and spiritual well-being. Originating as it did in Jamaica, and having spread first throughout the Caribbean region, Rastafari's *ital livit* is influenced by African food traditions that survived the Middle Passage and subsequent cultural adaptation, as well as the practicalities of local growing conditions and agricultural practices.[5]

While Rastafari has now attained worldwide reach, the original adherents to this tradition would have sought their *ital* food from among the plants growing in the Caribbean, both the indigenous and, like breadfruit, the creole. Growing abundantly, available year-round on many islands, and full of simple nutrition, breadfruit is a popular *ital* food item. Having been introduced into Jamaica more than a century before the emergence of the Rastafarian movement, breadfruit was already a creolized crop as this new creole religious movement began to spread. Today, in Jamaica, *ital* food may be found at establishments ranging from humble, traditional food stalls to flashy new restaurants with online reviews and social-media followings, capitalizing on the international appeal of both Rastafarian culture and local, plant-based cuisine. At both, diners are likely to find *ital* meals such as "rundown," a traditional dish of breadfruit stewed in coconut sauce and liberally seasoned with local herbs. A similar scene unfolds along the urban banks of the North River, running through Kingstown, the capital of St. Vincent and the Grenadines, where several brightly colored plywood shacks with "*Ital* Food" hand-painted on the side serve up roast breadfruit alongside callaloo, chickpeas, and root crops for just a few dollars per plate. One does not have to be Rastafarian to partake in the *livity*.

As illustrated by its acceptance into Rastafarian *ital* cuisine, breadfruit has undergone a cultural creolization in the Caribbean. It has developed some new qualities that distinguish it from *'ulu* in the Pacific, indicating adaptation to a new human environment. For example, older dishes like Grenada's oil-down and newer ones like Barbados's hollowed-out-and-filled roast breadfruit became popular in the Caribbean before anything similar was ever tasted in the Pacific, despite centuries of culinary experience with *'ulu*. Slicing breadfruit thin and frying it in oil appears to have been a Caribbean innovation as well.

New, creolized uses for the tree's other resources have also emerged. In the Caribbean, especially on the islands of Barbados and St. Vincent, a tradition arose to make coffins of breadfruit wood. This was normally done for the funerals of poorer people—breadfruit wood being both abundant and easy to work with—but when the governor general of St. Vincent and the Grenadines, Frederick Ballantyne, died in 2020, he was buried in a custom-built breadfruit wood casket, according to his wishes. A foreshadow of the South African Archbishop Desmond Tutu's 2021 burial in a "simple pine coffin with rope handles," in keeping with his wish to be buried in "the cheapest coffin and [without] any lavish funeral expense," this act of humility may have served to raise the status of the craft as the Vincentian public saw, for the first time, a revered government official being buried in a humble breadfruit casket. Breadfruit wood has long been used in the Pacific islands to construct buildings and objects such as boats, bowls, and surfboards. And while I have read about the practice of making funeral wreaths from the leaves of *'ulu* trees and, in preparation for burial, bodies being wrapped in *tapa*—the cloth sometimes made from breadfruit bark—I found no record of breadfruit wood being used to construct caskets on any Pacific island. Rather, it appears to be part of a cultural creolization of breadfruit in the Caribbean.[6]

Plants can creolize botanically as well as culturally. Through the "actions of reciprocity, the give and take with the land" described by Kimmerer, breadfruit in the Caribbean has changed ever slightly from what it was in the Pacific. Botanical creolization involves the kind of biological adaptations inherent in evolution by natural selection: the anatomical and physiological modifications that occur when a population of organisms finds itself in a novel environment. Traits that conferred an advantage in the old environment may not do so in the new, even as new traits emerge as advantageous in a plant or animal's new home. Over time, individuals of the same species living in different environments become distinguishable from one another.[7]

One of the first to recognize this concept of plant creolization was the eighteenth-century explorer Antonio de Ulloa, who described seeing in Peru "fruits . . . of the Creole kind, being European Fruits planted there, but which have undergone considerable alterations from the climate." Ulloa identified one of the key elements of a creole plant, and by extension, anything that's been creolized: the inherent differences it exhibits as a result of the different climate of its new home. To climate in the strictly meteorological sense, I would add other elements of the environment, including soils, microbial communities, and human preferences.[8]

The historical record bears witness to breadfruit's botanical creolization in the islands of the Caribbean, which seems to have been an easier and faster process than its cultural creolization there, at least at first. Breadfruit grows well in these tropical climates, these coralline and volcanic soils, but its acceptance by the people it was intended to feed has been slow and inconsistent. Before even departing Jamaica, Captain Bligh wrote to Joseph Banks claiming that "the Plants are thriving delightfully [and] we may therefore consider the Breadfruit already established in this Country." Christopher Smith, one of the botanists aboard the *Providence* tasked with tending to the plants during the voyage wrote to Banks after completing the mission, assuring him that "the Breadfruit & all the other Tahitian Plants are growing as luxuriant as I have ever seen them on the Plains of Matavai." James Wiles, who supervised the botanical gardens that had received breadfruit trees in Jamaica, wrote to Banks later that year, stating that the breadfruit was "thriving with the greatest luxuriance." Three years after the introduction to St. Vincent, Alexander Anderson wrote to Banks to report that, in the botanical garden he directed there, "the Bread fruit thrives (if possible) better than in its native soil," and that "all the trees in the Garden . . . have been bearing constantly for 18 months past." Finally, in 1801, Wiles wrote Banks again, this time concluding that breadfruit had become "perfectly naturalized" in Jamaica.[9]

Still, despite breadfruit's thriving botanically, the gardener Alexander Anderson would later report to Banks from St. Vincent about the strange case of "some people who under value such a valuable acquisition." Anderson explained, however, that "these are only some self conceited & prejudiced Creoles, and you well know that many despise those things, tho' useful, they have had no hand in obtaining." In Anderson's view, there were some people who had been born in the Caribbean islands, "Creoles," who turned up their noses at this newly arrived, yet-to-be-creolized crop. To him, it seemed they preferred instead other foods like plantains and yams, which had also been imported, but earlier and perhaps just earlier *enough* to seem as though they had always been there.[10]

The environment into which a person is born is that person's native environment. Until they expand their view through travel, or reading, it's all they've ever known; it's their baseline. But as the environment changes, baselines shift from generation to generation. In 1995, fisheries scientist Daniel Pauly introduced the concept of "shifting baselines," which he described as a situation in which "each generation of fisheries scientists accepts as a baseline the stock size and species composition that occurred at the beginning of their careers, and uses this to evaluate changes." Pauly went on to explain that "when the next generation starts its career, the stocks have further declined, but it is the stocks at that time that serve as a new baseline. The result obviously is a gradual shift of the baseline [and] a gradual accommodation of the creeping disappearance of resource species." Just as a fisheries scientist might subconsciously judge the "creeping disappearance" of fishes against the baseline accepted at the start of her own career, even if this baseline was itself the product of an altered system, someone who favors local foods, or is prejudiced against culinary introductions, might subconsciously accept as "local" all the foods that were present during his own childhood and reject those that came later. Even if those childhood foods had been, in fact, introduced, they would form part of the baseline against which newer imports would be judged.[11]

In this way, the concept of shifting baselines is useful to help understand why certain Caribbean food crops like plantain and yam, even though they had been introduced before—plantains in the early sixteenth century and yams several decades later—would have *always been there* in the minds of people born after their introduction. In the early 1790s, when breadfruit first arrived in the region, plantains and yams were already part of the Caribbean's botanical and culinary baseline. A food introduced within one's own lifetime, like breadfruit, might take some time to be accepted, to be creolized.[12]

Of course, creolization—of both people and plants—does not merely indicate transplantation. You don't creolize to every new country in which you travel. But if you stay long enough, your children might. Food historian Candice Goucher, in an essay on the Caribbean practice of eating iguanas, wrote that as the use of the word *creole* increased within the region during the late eighteenth and early nineteenth centuries, its definition expanded as well. By 1804, calling a person *creole* "tended to connote an attitude of adjustment to a slower, more leisurely lifestyle and access to greater material wealth, but also cultural accommodation of both African-derived influences and European expectations." In other words, creolization in the Caribbean came to include taking it easy, having access to more resources, and a blending of diverse cultural influences. A free creolized person in the Caribbean, then, might not mind languidly whiling away the time on their veranda while a breadfruit slowly roasts atop a coal pot and an iguana stews on the stove.[13]

By now, nearly two hundred and fifty years since its introduction to the region, breadfruit is part of every Caribbean person's baseline. The tree is a ubiquitous part of the landscape, and the fruit is a familiar presence in markets and in recipes on nearly every Caribbean island. But has it truly creolized in the region, botanically speaking, so that a breadfruit tree "born there" can be distinguished, genetically, from one that was "brought there?" Or has the process of cultural creolization, after getting off to a slow start, now outpaced breadfruit's botanical creolization?

In an effort to find out, agronomist Laura Roberts-Nkrumah has collected and analyzed specimens from breadfruit trees grown throughout the Caribbean region, searching for unique traits. The culmination of her life's work is the repository of breadfruit tissue samples housed at the University of the West Indies-St. Augustine, in Trinidad, and the book that catalogues and describes this collection. Roberts-Nkrumah explained to me by email that the effects of the unique Caribbean environments—as distinguished from those of the Pacific islands—along with the particular preferences and tastes that influence human-driven selection may have led to differences between Pacific and Caribbean breadfruit as seen in factors like fruit flavor and texture, seasonality, and tree growth. She explained that even slighter differences from island to island within the Caribbean can lead to variation among breadfruit grown throughout the region.[14]

What Roberts-Nkrumah described sounds a lot like the development of terroir: a distinguishable set of characteristics picked up from the environment of a particular place by the fruit or other food produced there. Terroir is the reason why bottles of wine or bags of whole-bean coffee proudly display the place where the grapes or the coffee cherries were grown. Tobacco's terroir—together with the scarcity resulting from the U.S. embargo—is behind the mystique of Cuban cigars. Cheese has terroir. Some gourmands say they can taste the specific flowers, grasses, and herbs upon which a cow

Figure 9.1 Clayton Ryan roasting breadfruits in Colonaire, St. Vincent and the Grenadines. (Photograph by Hannah-Marie Garcia.)

grazed before giving the milk that would be turned into their Emmentaler or Gruyère. To maintain the qualities of terroir against the diluting effects of imposters, the European Union has passed Protected Designation of Origin laws, which ensure the exclusive rights of local food producers to use geographically derived names for their products. Bubbly wine can be called *champagne* only if it comes from the Champagne region of France. The only olives legally called *kalamata* in Europe are from the Kalamata region on Greece's Peloponnese peninsula.[15]

The land of the Caribbean comprises islands large and small, high and low, volcanic and coralline. The mainland coasts of Central and South America are considered parts of the region as well and these include both the geologically young Isthmus of Panama and the coastal plain sloping upward toward the ancient Guiana Shield. Warm, moist trade winds blow in consistently from the Atlantic, giving nearly every Caribbean island a windward side and a leeward side, each readily distinguishable from the other, and creating diverse climate zones within each island. Biomes vary across the region with rainforests, arid savannas, and cool mountain peaks all represented. It may be possible, then, for someone who really knows their breadfruit to distinguish—by taste, texture, color, or scent—a fruit grown in the old limestone of Barbados from another fruit of the same cultivar, picked from a tree rooted in the young volcanic soil on the slopes of St. Vincent's La Soufrière. Even more, the qualities of a creolized Caribbean breadfruit would be distinguishable from one grown on a Pacific island like Hawai'i or Tahiti.

We can think of terroir as a step toward creolization and creolization as a step toward speciation. Coffee grown in the foggy Guatemalan highlands will, to someone trained to detect it, produce a different taste, a different experience, than coffee grown on one of Brazil's enormous plantations. Both are certainly of the same species, *Coffea arabica*, and could even be members of the same cultivar, 'Bourbon' or 'Typica,' for example. The differences in the cup are attributed to details too small for these taxonomic classifications to capture. But you can imagine the human-aided genetic divergence that takes place over time. The Guatemalan growers respond to consumer demand, selecting for propagation those plants that produce coffee cherries with desired Guatemalan characteristics. They also respond to the possibilities presented by their environment, selecting for propagation those coffee plants that grow the best in the volcanic soil of their high-elevation farms. The Brazilian growers do the same, but in response to different consumer preferences and different environmental conditions and having started with a different set of individual plants. Over time, Brazilian coffee and Guatemalan coffee diverge into separate cultivars; over more time, *much* more time, into separate species.

If this same effect were to be detected in breadfruit, you would be able to taste the difference between two fruits of the same cultivar—the ubiquitous 'Ma'afala,' for example—if they had been harvested from trees growing in different environments or on different islands. Greater variation would be seen in fruits taken from breadfruit trees of *different* cultivars since they are further along in the process. In this case, differences would have arisen longer ago and aided by human selection to propagate trees with certain specific, desirable characteristics. The selection process likely would have started when someone noticed a particular new trait in a tree or a fruit and decided to try to replicate it.

In species that reproduce sexually, both plants and animals, the copying and mixing of DNA from both parents to create the genes of the offspring sometimes includes transcription errors. Traits derived from these errors are called mutations. Evolutionary theory teaches us that most mutations are harmful, or neutral at best. Those rare mutations that confer some benefit to the organism may lead first to variation, then, with enough time, to speciation. But breadfruit is almost always propagated clonally: through air-layers, root shoots, or tissue culture. As plant geneticist Nyree Zerega explained to science writer Amanda Morris, "clonal propagation ensures that you get what you expect."[16]

Still, variation in clonally propagated crop species like breadfruit can occur, due to somatic mutations, or mutations not associated with reproduction. Rather than affecting the entire tree, a somatic mutation in breadfruit might give rise to a new trait found only in a single branch. Taking a root shoot, then, would not propagate the mutation, unless the mutation were found in the root shoot itself, but the mutated branch—or a smaller branch growing off it—can be air-layered and grown into a new tree. The new tree will then carry the new trait.

When Captain Bligh arrived in the Caribbean in 1793, he recorded in his log that HMS *Providence* carried five cultivars of breadfruit: four from Tahiti and one from Timor. Only a few additional cultivars have since been introduced into the region. After these initial introductions, somatic mutations—coupled with human-facilitated selection and propagation—have led to further diversification, entirely separate from whatever progress was occurring with these Caribbean trees' clonal parents, still growing half a world away in the Pacific. In 1793, then, when the *Providence* arrived at St. Vincent two broad lineages of breadfruit first came into existence. Two hundred and thirty years later, there was a family reunion of sorts.

In much the same way as people descended from immigrants sometimes travel to their ancestral homelands to meet relatives descended from those who stayed behind, scientists have been working recently to reconnect, through genetic analysis, the long-separated Caribbean and Pacific breadfruit lineages. In 1793, when Bligh's breadfruit plants arrived in St. Vincent and Jamaica, a new branch of the plant's family was established. This line would undergo its own evolution in response to its new environment. It would change according to the differing selective choices of its human propagators. New traits would develop.

Botanical researchers studying breadfruit in the Pacific and the Caribbean are interested in learning what differences might have emerged in breadfruit's DNA, essentially the first tentative steps at Pacific *'ulu* and Caribbean breadfruit becoming two plants, distinguishable from one other at a genetic level. To accomplish this goal, a team led by Lauren Audi, then a graduate student at Northwestern University, collected tissue samples from breadfruit trees grown in dozens of countries in the Caribbean and Pacific—along with samples taken from breadfruit growing in Asia, Africa, and Latin America—for genetic analysis.[17]

The researchers' objective was to match pairs of breadfruit cultivars found in both the Pacific and the Caribbean, to say this one and that one are the same. Remember, unless a somatic mutation has occurred, a root shoot or an air-layered branch is a clone of its parent tree. When a mutation does occur, it can lead to the development of a new cultivar if the mutated branch or shoot is allowed to propagate. Audi and her team knew that if they found a breadfruit cultivar in the Caribbean without any Pacific matches, that might indicate botanical creolization. This would most likely happen due to human selection. A grower would notice that a fruit with a new, desirable quality had emerged on his tree and would selectively propagate that particular tree to produce more of that desirable fruit.

Some of the diversity of breadfruit cultivars in the Caribbean may simply be a matter of naming. Audi cautioned her audience not to read too much into the long lists of cultivar names her team had compiled, since "multiple names may be applied to the same genotype, or multiple genotypes may be represented by a single name." Still, the researchers noted "the possibility of new variations arising . . . in St. Caribbean Islands." In other words, creole cultivars.[18]

In addition to the peer-reviewed academic publication reporting their findings, Audi's team also produced a brochure about breadfruit history and diversity in the Caribbean for distribution among the public within the region. In this brochure, freed somewhat from the strict autoskepticism required by the academy in scientific writing, Audi was able to speculate more broadly than in the journal article. For her Caribbean audience, she wrote that, "the time since breadfruit was introduced to [St. Vincent], over 200 years ago, may very well have been enough time for new cultivars to have arisen and be selectively grown."[19]

What might these new cultivars have been called? In an appendix to Audi's brochure, she included descriptions of several Caribbean breadfruit cultivars. For those that have been traced to their Pacific counterparts, she gave both names. 'Waterloo' in St. Vincent was matched with 'Paea' in Tahiti; 'Kashee' in St. Vincent is most likely the same as 'Puero' in Tahiti. A cultivar that is very popular in the kitchens of both the Caribbean and the Pacific is called 'Bligh' in St. Vincent, 'Cassava' in Jamaica, and 'Maire' in Tahiti. A particularly odd cultivar found in St. Vincent is called 'Sour sop' and may be matched to the breadfruit that the crew of the *Providence* collected during a stop at Timor along the way from Tahiti to the Caribbean. A nineteenth-century naturalist had described a "small and very inferior" cultivar then called 'Timor' growing in the St. Vincent Botanical Gardens. Audi and her research team found this cultivar to be very distinct from the others growing in St. Vincent and not often cultivated. It seems, if 'Sour sop' is indeed the cultivar that Captain Bligh brought from Timor, his efforts there did not pay off as well as the work done to collect breadfruit in Tahiti.[20]

For several Caribbean cultivars, a Pacific match was not determined. For these, Audi gave only the Caribbean names in the brochure. A group of unmatched cultivars all presciently included the term *creole* in their names: 'Creole Black,' 'Creole Common,' 'Creole Ready Roast,' and 'Creole Red.' Common names for plants arise from everyday usage and the reason for the application of the term "creole" to these breadfruit cultivars is, for now, lost to history. If further genetic research shows that these names indeed represent cultivars found only in the Caribbean and not matched to precise Pacific counterparts, it will turn out that it was serendipitously appropriate to refer to these particular Caribbean cultivars as *creole*: breadfruit varieties "born" in the Caribbean region.

CHAPTER X

Roast or Fry

Sandra Mason, in her role as governor general of Barbados, delivered the 2020 Throne Speech at the opening of that year's Barbadian Parliament in September. In keeping with the times, Mason began her remarks by discussing the COVID-19 pandemic, particularly the effects it had had on the economies of tourism-dependent countries like hers. This was a worldwide pandemic, she acknowledged, and with every country dealing with its own catastrophe, "no one is coming to rescue us." Rather, Mason said, "the solutions to the present circumstances are ours to find, ours to craft and ours to implement." Mason stayed on the topic of self-sufficiency for the remainder of the speech, invoking the legacy of Errol Barrow, her country's first prime minister, who had led Barbados to independence from Britain in 1966. This was a classic political address: the problem statement, the need to come together, the reference to a hero of the past.[1]

But the specific Barrow quote that Mason cited may have caught her audience by surprise. It seemed to peer into unfamiliar political territory. Barrow, Mason said, had cautioned in 1966 against "loitering on colonial premises" after securing independence. In late 2020, the government of Barbados was led by Prime Minister Mia Mottley, but, as a parliamentary constitutional monarchy, still retained Queen Elizabeth as the official head of state. Other former British colonies such as Australia, Canada, India, Jamaica, and Kenya have the same arrangement.[2]

On that day in Bridgetown, after noting that more than a half-century had passed since Barrow had peacefully led Barbados to independence, Mason affirmed the obvious fact that Barbados could then "be in no doubt about its capacity for self-governance" and that "the time has come to fully leave our colonial past behind." Her audience—the Parliament, the prime minister, the Barbadian public, and, indirectly, the queen—knew exactly what the governor general meant, though few had expected this announcement to come without a referendum and in the midst of a global pandemic. It was time to take what Mason called "the next logical step" in political independence by becoming a republic and replacing the queen with an elected president as head of state.[3]

This happened with remarkable speed. Barely a year after the Throne Speech, on November 30, 2021, at a ceremony that drew the attendance of both then-Prince Charles and pop star Rhianna, the transition was made: Mason took office as president, and Charles read the queen's message of goodwill. A few weeks later, Prime Minister Mottley called for a new election to be held in early 2022. For the first time in nearly four hundred years, Barbados would no longer recognize the authority of the British monarch.

What does it mean for a country to leave its colonial past behind? Surely it involves more than merely changing the governor general's title to president and having a septuagenarian prince deliver his mother's farewell note. What of the remnant structures of colonialism that remain in place, the distributions of wealth and power along racialized and genealogical lines? What of the colonially formed attitudes toward food, work, and the environment? What of the dependence upon the former metropole? "Who's going to come to our aid the next time we get hit by a bad hurricane?" a Barbadian friend, clearly opposed to his country's new status, asked me, "It's always been the UK."

In Barbados, perhaps more than any other former British colony, the cultural association with colonial power is strong. One mid-twentieth-century travel writer described Barbados as "a shire that had drifted loose from the coast of England and floated all the way to these tropic waters." Since being claimed for King James in 1625, Barbados never changed hands, unlike many of its neighboring Caribbean colonies, remaining continuously British until independence in 1966. Even then, the monarch remained head of state until the transition led by Mason and Mottley in 2021. British products are readily available in Barbadian shops. Afternoon tea can be had at several former plantation houses and upscale resorts. The Anglican Church is the largest denomination in the country. Cricket, horse racing, and polo are popular sports. The country's nicknames include "Bimshire" and "Little England."[4]

Breadfruit is a creole food crop in the Caribbean: local, yet introduced; introduced, yet established. It is nutritious, plentiful, and ubiquitous in the region. At the same time, it is a relic of a colonial past when people, plants, and commodities were transferred around the world all for the benefit of wealthy, industrialized nations, and of the Atlantic slave trade—a systemic injustice equaled in history only by genocides and similarly scaled atrocities. To some people in the Caribbean, the sight of a breadfruit tree, or the taste of breadfruit in a meal, recalls memories of slavery and colonization. To some, breadfruit still carries a stigma. Can Sandra Mason's call to move fully beyond Barbados's colonial past include a reimagination of breadfruit in the Caribbean?

The creolization of breadfruit—both botanical and cultural—suggests that it can indeed be reimagined. The Barbadian cookbook author Austin Clarke has written that "from a bad, disreputable journey, a segment of the Middle Passage, a good thing spring up, a green and large harvest of breadfruits." Could it be that something as simple as a breadfruit, roasted atop a Caribbean coal pot or wood fire, might be reimagined to embody both the cruel past of Caribbean slavery and the bright future of "a green and large harvest?" For years now, I have been searching the islands for evidence of this reimagination.[5]

One morning, I visited the Barbados Agricultural Development and Marketing Corporation (BADMC), an organization established by the government in 1993 to support Barbadian agriculture and to develop and market innovative food products for sale both locally and abroad. According to its own Government Information Service, as recently as 2019, Barbados annually imported nearly US$250 million worth of food. This in a country that traces its history back to a time when its sole colonially designated purpose was sugarcane agriculture. Sugar has been declining in Barbados for decades and the BADMC's mandate could be viewed as a way to respect that agricultural heritage while steering the country toward a diversified, nutritious, self-sufficient, and profitable future of food production.[6]

I had come to BADMC to meet Sonia Blackman-Francis, manager of the corporation's Food Innovation Unit, and to find out whether this future was within reach. Sonia and I donned hairnets and walked inside the small room where one of the organization's flagship products, Carmeta's Fine Breadfruit Flour, is made. At a stainless-steel table sat two aproned and hairnetted women, each with an enormous plastic bucket of sweet potatoes by her side, peeling. It turned out I had visited on sweet potato day, not breadfruit day, but Sonia assured me the process was identical.

At this small food-production facility, Sonia introduced me to the sweet potato peelers—I tried very hard to imagine they were actually peeling breadfruits—and walked me through the process of making flour. The sweet potatoes (that is to say, the *breadfruits*) were peeled and washed and then put through the culinary equivalent of a woodchipper to produce small, irregularly sized slices. These slices were then laid out on trays and stacked into a commercial food dehydrator until they lost nearly all their moisture. After drying, the slices were milled into a fine, gluten-free breadfruit flour that was sealed into the BADMC's signature Carmeta's-brand minimalistic-style packaging and sent out to the shops.

We retired to Sonia's air-conditioned office to discuss breadfruit-based food production in Barbados. My first question was an obvious one: who is Carmeta? The answer was not so simple, only because Carmeta Fraser was such a complex figure in Barbadian food history. A tireless advocate for the role of self-sufficiency in food security, Fraser was known for her down-to-earth approach. She taught classes on food preservation techniques, hosted cooking programs on television, wrote cookbooks with recipes that relied heavily upon Barbados-grown ingredients, and spun simple maxims like, "Eat what you grow; grow what you eat." She also maintained a small farm and restaurant, an early precursor to today's "farm to table" movement. Fraser was interested in shifting the dietary habits of Barbadians toward more homegrown foods, not only to strengthen her country's agricultural economic sector, but because she believed that local foods simply taste better and, generally, are more nutritious than imports. As early as 1993 she called "non-sugar agriculture" the "unsung hero" of the Barbadian economy, touting how even then, when the sugar industry was stronger than it is today, other crops brought in nearly four times as much foreign revenue.[7]

Sonia went on to explain the difficulties of breadfruit flour production. Yes, the trees are plentiful, but many have gone unpruned for so long that picking breadfruits from the highest branches is an act of self-endangerment. I knew the risks of harvesting breadfruit from tall trees well. When I first began to dive into the literature on breadfruit during the pandemic, I set up a Google alert for any new instances of the word "breadfruit" published online. Most of the results I get are articles from Caribbean news sources where a common theme—I probably see it in an alert every other month or so—is a breadfruit harvester being injured or killed, having climbed too high and fallen or, a frighteningly common occurrence, being electrocuted by touching powerlines that pass through the branches of an unpruned breadfruit tree. One news story from St. Lucia reported three separate breadfruit-related deaths there, all within a single week. A resident told the journalist, "It seems like these days breadfruit trees have a jinx." The BADMC, in an effort to keep would-be breadfruit harvesters safe, rarely buys from pickers who are known to climb to unsafe heights or from parts of the island where the trees grow tall and unpruned. With a population of less than 300,000, Barbados is a small enough society to know the usual suspects when it comes to unsafe breadfruit harvesting practices.[8]

Marketing, too, has its challenges. According to Sonia, many Barbadians would rather make a quick stop at Chefette, a local fast-food chain, than take the time to prepare a home-cooked meal. "Of course, if you give them a roast breadfruit they'll eat it," Sonia chuckled. The problem lies with the willingness to attempt cooking, especially with novel ingredients like breadfruit flour. To this end, the BADMC has produced a series of recipe cards, each of which gives simple, step-by-step instructions for dishes that feature Carmeta's-branded products. Sonia gave me cards with recipes for breadfruit fish cakes and breadfruit great cake, the latter of which, in addition to breadfruit flour, calls for twelve ounces of stout, two cups of rum, and a cup of brandy. Maybe *that* will convince a few culinary novices to give cooking a try.

I left the BADMC with my head full of new information but my stomach still empty of breadfruit. To remedy this inequity, I drove to Bridgetown's working-class Black Rock neighborhood, where I found a small takeout food shop called Yelluh Meat. Described on its website as "a street-food ambrosia experience inspired by traditional roasted breadfruit," the dishes served at Yelluh Meat are a remarkable mix of old and new attitudes toward Caribbean food.

The place consists of an outdoor cash register set up on an open patio in front of a kitchen. Breadfruits are roasted across the parking lot on an open fire until the skin is thoroughly blackened, then halved. The fruit's flesh—the eponymous "yellow meat"—is scooped out, diced, and mixed with whichever fillings the customer has ordered before being replaced into the hemispheric natural bowl of the breadfruit rind.

Black Rock doesn't see many tourists. It's dense and urban, with lots of concrete and traffic, and not particularly convenient to any of Barbados's beaches. Nearly everyone queued at Yelluh Meat was local. The meals are reasonably priced and Yelluh Meat's menu features flavors that appeal to a local Barbadian clientele. I opted for the vegetarian filling, which included black beans, sweet corn, diced peppers, mango, pineapple, and a generous douse of garlic butter. Other options included such varied fillings as pickled herring, pig tails, or saltfish buljol.

Before my visit to Yelluh Meat, I had asked Dwight Forde, who owns the restaurant with his partner, Kim Hamblin, when would be most convenient for me to stop by not just for a meal but to talk about their operation. After the lunchtime rush, he told me, around two in the afternoon. I arrived closer to three and still, a line of customers snaked out from under the covered pavilion, people enduring the tropical sun while they waited to place their orders. I joined the queue and chatted with a few of the other customers. One couple stood out among the Barbadian locals: a mother and daughter on vacation from Chicago who told me they had "paid the taxi driver $60 to get here for the breadfruit."

After putting in my order with Kim, who was working the register that day, I walked across the parking lot to find Dwight busy breaking down wooden pallets to fuel the breadfruit-roasting fire. He took time to talk with me while he worked.

Dwight and Kim met in a youth entrepreneurship program at the University of the West Indies. The design for a breadfruit-based restaurant was their final class project. They had been guided by the question, "How do we provide the most true, authentic experience of our culture and heritage?" Dwight explained that, while breadfruit is roasted throughout the Caribbean, eating it while still in the skin is unique to Barbados. Elsewhere in the region, after roasting, the blackened skin of the breadfruit is scraped off and the cooked flesh is sliced to be served. Barbadians have traditionally cut their roasted breadfruits in half and scooped the flesh out to eat. Dwight and Kim's innovation, he said, was to "change the perception of breadfruit" among their future clientele.

He took a moment to cut a pallet into pieces with an electric saw and then continued: "In some parts of the island breadfruit was regarded as poor man's food or even pig food. There was a low value to breadfruit." But,

Figure 10.1 Dwight Forde preparing a breadfruit for roasting at Yelluh Meat in Black Rock, Barbados. (Photograph by Russell Fielding.)

Dwight insisted, breadfruit's low value was forgotten in times of need. "If you had a breadfruit tree that meant your household didn't go hungry." Dwight and Kim's main innovation was to show Barbadians that breadfruit could have high value. "People didn't understand that you could sell breadfruit," he told me, "They didn't know it had *monetary* value."

Now, when Barbadians see Dwight, Kim, and their staff wearing crisp, black polo shirts with the Yelluh Meat logo, or when they follow the company's well-curated social media accounts, it's apparent that breadfruit's value is more than just hunger prevention. Breadfruit is supporting a profitable business, the most successful outcome of Dwight and Kim's youth entrepreneurship class.

And the food is wildly popular. Later I talked with Chris Alleyne of the Walkers Institute for Regenerative Research and Design, a Barbados-based environmental education organization, who told me about the institute's recent holiday party. Yelluh Meat had catered the event and "you had grown adults crashing the party to get at those breadfruit bowls," Chris told me. "Why are they so popular?" I asked. Chris explained the appeal through his own life history. He had grown up on the east coast of Barbados in the town of Bathsheba, a poor area, economically, but a place where breadfruit trees are plentiful. As a child, Chris would walk home from school each day down a route known as Breadfruit Alley. He and his friends would pick their supper along the way. Upon reaching home, the breadfruits would often be roasted in a communal fire, with many families eating together and sharing their meals.

Chris did well for himself in terms of education and career, but his upward progress meant a degree of separation from the simple, community food traditions of his past. "People like me," Chris explained, "we grow up and get jobs, move to the 'heights and terraces,'" a term used to refer to the well-to-do neighborhoods of Barbados. "We long for that community. Roasting breadfruit brings people back together." To Chris and many others in Barbados, Yelluh Meat's breadfruit bowls do not represent an unwelcome step toward commercialization, but a convenient way to recapture the community spirit they have nearly lost.

Indeed, many people I met throughout the Caribbean would speak of breadfruit not only as a food source but also as a nexus of neighborhood life. Like the villagers in the fable "Stone Soup," a Caribbean community rich in breadfruit often has a rich tradition of sharing as well. One neighbor brings breadfruits from his tree, another brings herbs from her garden, while another still brings fish from her brother's boat. With everyone's humble contributions put together, the community feasts.[9]

Finally Kim called out my name. My order was ready. I retrieved a compostable grocery bag tied around an aluminum foil-wrapped half-breadfruit. Since Yelluh Meat only offers a few picnic tables, most in direct sunlight, I drove down to Pebbles Beach on Carlisle Bay, where I knew I could sit in the shade at the Barbados Cruising Club and enjoy my late lunch.

The breadfruit that had been scooped out and diced was mixed with the ingredients I'd chosen and sauteed to a gentle crisp. The breadfruit that still formed the rim of the bowl was softer, chewier, and bore a distinct woodsmoke flavor from the fire. I was glad to have asked for garlic butter, which I could taste in every bite. The meal was delicious, and I scraped the bowl down to the charred exterior with my spoon just to make sure I got all the soft, smoky breadfruit. The Cruising Club's al fresco restaurant was open, and I could see several tables occupied by diners enjoying a late lunch. My family and I had eaten here many times, and I knew the menu well: grilled fish, hamburgers, french fries—all familiar foods to locals and tourists alike. I wondered what it would take for an old-and-new creation like Yelluh Meat's breadfruit bowl to be part of a broader and better-known reimagination of Caribbean cuisine, for the tourists lounging on Pebbles Beach to know this food, to have access to it without having to shell out for a $60 taxi ride. Could other young, creative culinary entrepreneurs in the Caribbean like Dwight and Kim see breadfruit through the process of leaving its colonial past behind, just as their islands have been doing in the stages of emancipation, independence, and full political autonomy?

To learn more about breadfruit's emergence from colonialism as a food of independence, I visited other Caribbean islands to see how it was being used in both traditional and nontraditional ways. I found an interesting connection on several islands between new uses of breadfruit and hurricanes, particularly the devastating Category 5 storms of the 2017 season, Irma and Maria. On St. Barthélemy, I met Charles Moreau, who explained that in addition to his day job as a museum curator, he was "an artist too, unfortunately." I puzzled over Charles's use of the word *unfortunately* as he handed me his phone so I could scroll through images of several of his charcoal sketches. Most were botanical, and I paused to look at a sketch of a large, round breadfruit with a few leaves attached to the stem. "Oh, I love *fruit à pain*!" Charles said, picking up on my interest. "After the hurricane," meaning Irma, "this tree fed us." Even on tony St. Barts, where luxury foods are imported from around the world and served to wealthy vacationers aboard yachts or in chic restaurants, trade networks are precarious and vulnerable to disruption. But the *fruit à pain* is there when nothing else is coming in. Island-hopping among the tightly clustered French-, Spanish-, and English-speaking islands of the northeastern Caribbean, I would hear similar stories again and again: the disastrous effects of a hurricane forcing local islanders to get creative in sourcing their food, especially when normal supply lines were broken.[10]

On a mountaintop outside the village of Jayuya in Puerto Rico, I found Marisol Villalobos and Jesús Martes at their farm and food-production company called Amasar. The term *amasar* means "to knead," hinting at the company's main products: breadfruit flour-based baking mixes. One wall of the main building was covered by a mural depicting a world map with the route of HMS *Providence* from England to Tahiti to Jamaica—the route by which breadfruit first came to the Caribbean—shown as a dashed line.

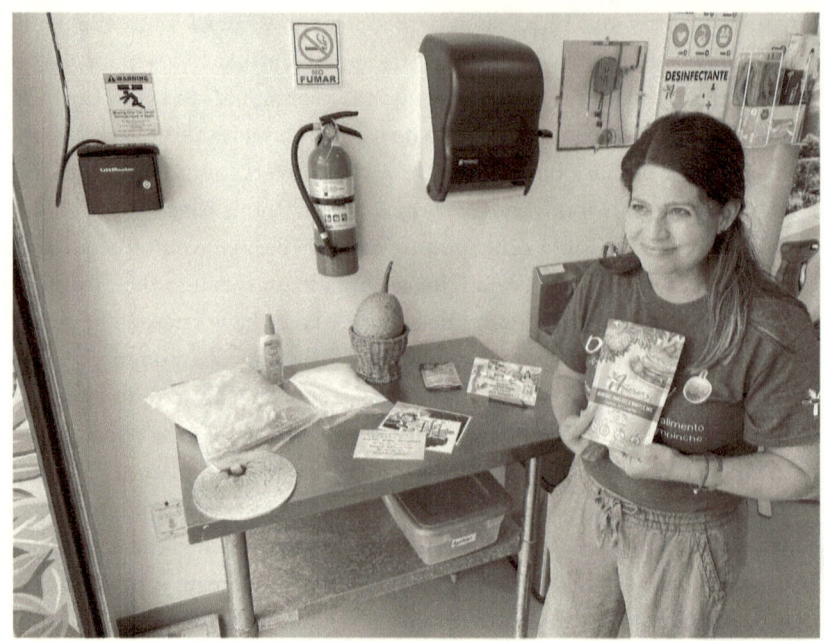

Figure 10.2 Marisol Villalobos with a package of breadfruit pancake and waffle mix at Amasar in Jayuya, Puerto Rico. (Photograph by Russell Fielding.)

Jesús left to prepare a meal for us while Marisol showed me the production facility where Amasar's breadfruit flour is made. During my tour, Marisol—who comes across like an enthusiastic elementary-school science teacher—laid out the history of breadfruit in Puerto Rico.

After the *Providence* delivered its cargo to St. Vincent and Jamaica in 1793, gardeners began distributing breadfruit throughout the Caribbean region, including Puerto Rico. There, breadfruit was gratefully received, especially by the *jíbaros*, a word Marisol translated for me as "hillbillies." Each *jíbaro* family, she explained, would plant two trees in front of their simple house: a breadfruit and a mango. "Dinner and dessert!" Marisol laughed.

Breadfruit-based dishes are common in Puerto Rican cuisine, though mostly with breadfruit standing in as a replacement for another, more standard starch like plantains or cassava. In these alternatives, *"de pana"* is usually added to the name of the dish: *mofongo de pana*, *pastelón de pana*, *tostones de pana*. *Pana*, in Puerto Rican Spanish, is one of several terms for breadfruit. The word is also slang for "friend," and somewhat nonsensical breadfruit-friend puns are frequently made. "Let's be *panas!*" declares the packaging of one of Amasar's products. Another Puerto Rican company called Panawest uses the hashtag *#todossomospanas* on its social media posts: We're all friends and we're all breadfruit.

Marisol and I walked around to meet Jesús on the shaded side of their house, where we shared a late breakfast of breadfruit-flour waffles, sliced bananas, and Puerto Rican coffee. There, we talked about their company and their dreams. Founded in 2016, Amasar had barely existed for a year when Hurricane Maria hit. In the immediate aftermath, Marisol said, "we had nothing but dried breadfruit and we didn't miss a pound!" Their startup food-production company quickly became an emergency food distribution center, with bags of breadfruit flour being given to neighbors that had lost everything. It was a lesson in self-sufficiency. "Puerto Rico imports 90 percent of our food," Jesús explained. "We have to change that."

To help make Puerto Rico more food-secure, Amasar has gone all-in on breadfruit flour, which Marisol calls *oro molido*, or ground gold. She and Jesús have planted dozens of trees on their own mountaintop farm and have donated trees to other growers. They have a special affection for the Future Farmers of America, having first met at one of the organization's events during high school. Marisol and Jesús have since donated breadfruit trees to fill out an acre of land adjacent to a local school and are now buying the breadfruits that the trees produce directly from the FFA students who tend them. And while Amasar doesn't sell pure flour alone, every one of its products includes breadfruit flour as a main ingredient. Most important, in terms of food security, Amasar sells mainly to the local Puerto Rican market. "In Puerto Rico, we want to build up the *pana* culture," Marisol told me.

Amasar may not stay so geographically focused for long. International organizations are starting to take notice of their products. From a shelf in her office, Marisol retrieved a small wooden trophy; its figurine looked like an Oscar but wore a chef's hat and held a covered dish. "This is my sofi," Marisol beamed. In 2022, Amasar's breadfruit pancake and waffle mix won a sofi Award for the best new breakfast product from the Specialty Food Association, a nonprofit trade association based in New York but with international reach. Then, the following year, Amasar won a "golden ticket" award from KeHE, an Illinois-based international distributor of specialty grocery products. While the sofi Award is more of a recognition of past accomplishments, the golden ticket is intended to pave the way for future sales of Amasar's products in grocery stores far beyond Puerto Rico.[11]

Jesús and Marisol introduced me to other Puerto Rican *panas de pana*, friends of breadfruit, whom I would meet in rapid succession. One evening, as the sun was setting, I arrived at a park in the small town of Caguas, about a half-hour's drive south of San Juan, where several food trucks had been congregating for more than a year and a few vendors had built permanent structures. One of these permanent food kiosks was a sharp-lined black building with two potted breadfruit trees framing the ordering window. A lighted white and green sign above the window gave the name: Mi Panapén (my breadfruit).[12]

Inside the stainless-steel kitchen I found the chefs and co-owners of Mi Panapén, Liannette Bezares and Edgardo Matías. There was a line of customers waiting to pick up their food and the chefs were moving quickly so I ordered a pineapple juice from the bar next door and waited in the common seating area for things to slow down. A family sitting next to me had just gotten their meal and I noticed the father eating tacos with the shells made from thin slices of breadfruit that had been fried and folded like a crispy corn tortilla. Edgardo and Liannette soon joined me at the table, along with their friend and part-time employee Ramfis Fuentes. Edgardo carried a basket of breadfruit chips and little bowl of salsa, which we devoured quickly as we talked.

I asked what kind of foods Edgardo and Liannette make from breadfruit and was surprised by the long list I got in response. Mi Panapén has a rotating menu that, on the night I visited, included tacos and pizza. The crust of the latter is made not from breadfruit flour, but from whole mashed breadfruit, to which a little wheat flour is added before being cooked atop a griddle. Other menu items include breadfruit pasta and *mofongo de panapén*, a breadfruit-based take on Puerto Rico's famous dish that's usually made from mashed plantains. For dessert, they offer breadfruit ice cream and a breadfruit-based crème brûlée. Edgardo told me, "I want to show the world we can do anything with breadfruit." With more than twenty recipes in their repertoire, Mi Panapén is making that point quite well.

Liannette scrolled on her phone to find a photo of their old food truck. It was expertly painted and looked like a mobile version of their current storefront. For about four years they operated the food truck, with some of the same breadfruit recipes that Mi Panapén serves today. When Hurricane Maria struck in 2017, the effects were devastating. The storm killed nearly three thousand people and caused tens of billions of dollars' worth of damage in Puerto Rico alone. Electricity was cut off throughout most of the island for months. Maria wiped out Puerto Rico's breadfruit crop.[13]

"We sold rice and beans for a year, waiting for the *pana* to grow," Liannette told me. Eventually the trees began to bear and the kitchen at Mi Panapén was once again filled with its namesake ingredient. As part of their plan for resilience in the face of future hurricanes, Edgardo and Liannette have leased land on which they now grow their own breadfruit trees. As they did before the hurricane, they also maintain relationships with more than a hundred growers, mostly just households with one or two breadfruit trees in their yards, going *casa a casa* to procure breadfruits, often just a few at a time. "Breadfruit is resilient," Edgardo told me, "like Puerto Ricans."

Not every example of Caribbean breadfruit redemption is to be found in the Caribbean. At a small restaurant in the trendy Buckhead neighborhood of Atlanta, I met up with Javion Blake, an engineer and entrepreneur from Jamaica who runs Jus Chill International, a food company that sells breadfruit chips and baking mixes made with breadfruit flour online, delivering to all fifty U.S. states. Jay, as he's called, sources his breadfruit from Jamaica and sees his work as more than just business. He's something of a self-appointed Caribbean cultural ambassador. "I have the opportunity to sell someone something and the opportunity to teach someone something," he told me. He wants to reassert the intrinsic value of Caribbean culture, both abroad and at home. "Colonized countries tend to take direction from outside sources," Jay explained to me. In Jamaica, he witnessed breadfruit's stigma as a poor man's food and as a reminder of slavery, but he wants Jamaicans to see that it doesn't have to stay fixed in that role. He explained how breadfruit is appreciated in Jamaica as a humble meal—"growing up, eating ackee and saltfish—you roast a breadfruit too and your breakfast is good to go!"—but that seeing it as a product desired by people in other countries adds to its local appeal, gives it economic value where once its only value was low-cultural and high-nutritional. "The stigma changes quickly if something is desired in the U.S.," he explained.

"In this business you have to tell stories," Jay said. To his American customers, the stories are printed right on the back of each package of breadfruit chips, recalling Jay's childhood, "growing up in the hills overlooking the city of Montego Bay," and his "memories of warm sunny days with blue skies and white-sand beaches." When American customers buy his products, Jay hopes, they're getting more than just a snack; they're actively participating in the transformation of breadfruit into an economically valued Caribbean export; three 1.2-ounce packages of the chips sell for $11.99 online.

To the Jamaicans working at Jus Chill's commercial kitchen, those growing, harvesting, and delivering the breadfruits, or even just those inspired by his success, Jay's is a story of redemption: a young man emigrating to build a lucrative business, creating and fulfilling a desire for breadfruit, that humble poor man's food now selling in its value-added state for more than three dollars an ounce.

Pericles Maillis is witnessing a similar transformation in the Bahamas, where he grows breadfruit and other tropical fruits on his large, coastal farm. "This land is solid rock," Pericles told me as we drove onto his property. Indeed, I could see bare limestone outcrops emerging between his rows of fruit trees. "We dig out potholes down to the water table and plant the trees in them. Sometimes we dig trenches so we can plant rows. I put a mango tree at every intersection," he explained. As we drove into a small grove of pothole-planted breadfruit trees, Pericles stopped the truck and summarized his operation: "These trees are grown with love."

As his name might suggest, Pericles, a descendant of Greek immigrants who came to the Bahamas as sponge divers during the late nineteenth century, has maintained his Hellenic identity even in this tropical environment. He explained, referencing the potential to mill breadfruit into flour, that "the best thing about breadfruit is its preservability. It's like Greek olive oil—you're making food last." We drove toward Nassau, and he pointed out the whitewashed Greek Orthodox church where his family worships. With the electric-blue ocean just behind, the church looked like a small piece of Santorini transplanted across the Atlantic. Along the way, we passed through Grants Town, a site where, following the end of slavery in the British Empire, formerly enslaved Africans—or those still being shipped to the Americas when abolition was enacted—were freed and settled. "Backyard breadfruit trees are plentiful here," Pericles said as we navigated Grants Town's potholed streets.

I asked him about the association of breadfruit with slavery in the Bahamas. "Oh, in my own mind it was slave food," Pericles said, reflecting on the opinions of his youth. But things are changing. "This is a real revival," he said, pointing at a midsized tree, heavy with fruit. He explained that breadfruit had become popular in the Bahamas during the early twentieth century with the arrival of skilled immigrants from the Caribbean islands—technically, the Bahamas are in the Atlantic, not the Caribbean. These Caribbean Islanders brought a love for breadfruit, which was already growing, though neglected, in the Bahamas, and taught the locals their own ways of preparing it. They also planted more trees. When mass tourism began during the 1960s, however, breadfruit fell out of fashion, being too much associated with the agrarian past of the islands, and not enough with their promising, lucrative, international future.

In the Bahamas, breadfruit's stigma is being overcome because of a recognition of its utility during times when self-sufficiency is needed. Hurricanes, pandemics, any disruption to the importation of foods from overseas or even just from other islands might force a community to feed itself from its own produce. How much demand is there for breadfruit in the Bahamas today? Enough to encourage illicit activity. Pericles cited one close-to-home example. Stopping his truck next to a gaping hole in the ground he lamented, "They stole a seven-foot tree from my farm!"

Island-hopping through the Caribbean, I saw that breadfruit was still somewhat marginal and had a lot of room to grow. To see the extent of breadfruit's flourishing in the region, I had to go back to the place it was first introduced: St. Vincent. "You need to talk with Michael Gloster, the minister of breadfruit," the man sitting across from me said. I was in Kingstown, the capital of St. Vincent and the Grenadines, talking with Jerrol Thompson, a local parliamentarian. "Your country has a minister of breadfruit?" I asked incredulously. "Well, that's not his official title," Jerrold conceded.

Officially an agricultural officer with the Ministry of Agriculture, Michael Gloster seems more at home in the fields and forests of the rural parts of his island than in a ministerial building in the capital. Short dreadlocks poked out from beneath his cowboy hat as we sat on the veranda of my hotel's lobby to talk. "Me and breadfruit, boy, it's a long story," Michael began. His discourse ranged between agronomy and history, taking in the technical aspects of pruning and the precise distribution patterns that followed Captain Bligh's original introduction of breadfruit to the St. Vincent Botanical Gardens in 1793. "There's more to the breadfruit story," Michael said, "too much ever to be told."

"I grew up with breadfruit, went to UWI on breadfruit," Michael said, referring to his studies at the University of the West Indies in Trinidad. After finishing his undergraduate degree, Michael stayed at UWI to complete his master's degree in horticulture, with a research focus on breadfruit. "The MPhil normally takes two years," Michael told me, "but longer when you study trees. I learned that the hard way. I had to propagate, plant, and wait for the trees to grow." Passing the time while his research subjects grew into mature trees, Michael had a long opportunity to think about the state of breadfruit in the Caribbean, particularly in his home country, and to imagine what could be.

Now the "minister of breadfruit," wants to see St. Vincent and the Grenadines increase breadfruit production to a commercial scale. He wants more value-added breadfruit products on the shelves of Vincentian supermarkets and packed into shipping containers for export. He wants to see what he calls "an Indigenous restaurant" in Kingstown serving local Vincentian food. "We should have a breadfruit brand; we should have a breadfruit industry," Michael told me. He described the Jamaican breadfruit industry as unworthy of the respect it has garnered within the region: "It came here first, then it went to Jamaica. Captain Bligh came here, spent a while, then he went to Jamaica. We need to claim breadfruit."

One of Michael's proudest achievements was the breadfruit festival he established with the intent to bring breadfruit out of the background and to the attention of the Vincentian people. "The things you have in front of you sometimes you don't see," he explained. It was a dispersed event, with activities celebrating breadfruit held at sites all around the island. Today breadfruit festivals are held in several other Caribbean and Pacific islands, but Michael says his was the first in the world. It's been paused since the one-two punch of the COVID-19 pandemic and the 2021 eruption of St. Vincent's La Soufrière volcano. But Michael has hope that the festival will return. Considering how breadfruit's importance exploded in the northeastern Caribbean following Hurricanes Irma and Maria, I wouldn't be surprised if the disasters in St. Vincent prove to have had a similar effect there.

He told me that he also wants to see breadfruit take on a larger role in the way St. Vincent presents itself to the world. Michael remembers the original flag of his country—the one with the breadfruit leaf—and wishes it would come back. St. Vincent and the Grenadines' current flag consists of vertical blue, gold, and green bands with three green diamonds in the shape of a V in the center. It was commissioned and designed in 1985 as a replacement for the breadfruit-leaf flag, which was seen as too complex, mainly due to the coat of arms, not the leaf.[14]

Radio host Roman Mars brought vexillology—the study of flags—to the attention of much of the public through his widely viewed 2015 TED talk on the subject. "Few things give me greater joy," Mars said during his talk, "than a well-designed flag." With this he displayed a large image of the Canadian flag on the screen, bringing cheers of patriotic pride from his Vancouver audience.[15]

Michael and I discussed the Canadian flag: the instant recognizability of its red maple leaf, its clean, crisp design able to be drawn, or at least approximated, even by a child with a crayon. St. Vincent and the Grenadines could keep its blue, gold, and green bands, Michael suggested, but replace the diamonds with a simplified, stylized breadfruit leaf. Its basic design, then, would resemble the Canadian flag but with the Vincentian color scheme and a leaf representing the tropical fruit tree so deeply associated with St. Vincent. I sketched a quick approximation in my notebook of the flag Michael was describing. "That's it!" he said.

Perhaps the best way to see breadfruit's redemption in the Caribbean is not on the dinner table or in the food truck—or even on a country's flag—but in a music video. In 2018, Jamaican dancehall artist Radion Tashaman Beckford, who goes by the stage name Chi Ching Ching, released an album titled *Turning Tables* that included a short, two-minute track called "Roast or Fry (Breadfruit)." In the video that accompanies the song, Chi Ching Ching, dressed in a chef's hat and apron, introduces a new hip-shaking, knee-wobbling dance as he holds aloft a whole roasted breadfruit in one hand and a skewered slice of fried breadfruit in the other.[16]

"You waan get yuh breadfruit?" he sings in Jamaican Patois, "Jus tell me if you waan breadfruit!" The video's scenes cut quickly from a Jamaican fruit market to an empty lot where Chi Ching Ching and friends stoke a fire and oil a pan, preparing breadfruits to roast and fry. All throughout, the song's chorus, "Roast or Fry!" rings out again and again like a chant while everyone dances. The last time I checked, the version of the video on Chi Ching Ching's official YouTube channel had more than four million views.[17]

Breadfruit's Caribbean history is a story of profound cultural change. It began as a novel import, clamored for by planters and detested by the enslaved laborers it was intended to feed. Since emancipation breadfruit has slowly creolized in the region, to the point that many Caribbean Islanders consider it to be a native crop, paying little to no attention to the story of Captain Bligh, the mutiny on the *Bounty*, or the tree's Pacific origin. Still, it has retained in some island societies a stigma: close association with poverty and guilt by association with the brutal injustices of slavery. Recently, local Caribbean chefs, entrepreneurs, farmers, and scientists have begun to recognize breadfruit's potential, establishing profitable and popular businesses based on the crop. And now, with its entrance into the youth-dominated culture of popular music, breadfruit seems to have finally and fully arrived on the Caribbean scene.

It is a testament to both the graciousness and the enterprising nature of the Caribbean people that a once-detested plant like breadfruit could be reevaluated, absolved of its role as an accomplice to slavery, and reclaimed as a resource for local food sovereignty, a source of both domestic and international income, and a representative element of the shared Caribbean culture. When I see breadfruit for sale at a Caribbean produce market, on a menu at a Caribbean restaurant, as a commodity bringing success to an expatriate Caribbean businessman, or in the lyrics to a dancehall hit, what I really see is the resilience of the human spirit. A spirit that says, "I will take what was foisted upon my ancestors against their will and I'll use it to advance my own cause. I will cook with it to attract diners to my restaurant. I will grind it into flour and sell it to customers who want gluten-free pancakes, by which I'll profit. And I'll even sing about it to a beat that makes you want to join the party."

I have conducted research in the Caribbean islands for more than twenty years. In that time I have come to believe that it just might be the most optimistic region on earth. And the story of Caribbean breadfruit exemplifies that optimism better than any other story I know. The farmers, produce vendors, chefs, business owners, and musicians of the Caribbean have turned something dark and mutinous into something bright and profitable. And they've even got a dance to go along with it.

PART III

The World

Map 3 The world, with relevant places labeled. (Cartography by Alison DeGr Ollivierre, Tombolo Maps & Design.)

CHAPTER XI

The Second-Best Time to Plant a Tree

Breadfruit is going global. In 2010, a Ghanaian pastor named Samuel Glory Abbey had a vision, which he described to me as we stood in the walled courtyard of his church in Accra, under the shade of a large breadfruit tree. In the vision, a woman led Samuel out into the forest and stopped in front of a tree that looked unlike any he had seen before. She showed him the tree's fruit: large, the size of a man's head, and bright green. "My son," the woman told Samuel, "this is the food of the future." He came out of the vision convinced that he had received a cryptic message from God and feeling an intense desire to find out more.

At the following Sunday's church service, Samuel described the vision in his sermon, ending with a plea to his congregation: "Does anyone know what this fruit might have been?" A woman stood up. "That's breadfruit," she announced, "poor man's food." The woman was from nearby Cameroon, where French colonialists had introduced breadfruit decades earlier and, in her estimation, it had acquired a reputation similar to what it had in some Caribbean societies. Samuel took the new information with gratitude but left the opinion behind. His vision had told him that this was the food of the future. He believed it could become everyone's food: rich and poor alike.

Samuel began meeting with agricultural scientists and extension officers to learn more about breadfruit. He saw his first breadfruit tree growing at a research station—its large, lobed leaves matched those of the tree he had seen in his vision—but time and again he was told the same thing: people here in Ghana don't want to grow it or eat it. Breadfruit had first been introduced into West Africa from the Caribbean during the mid-nineteenth century. The explorer Richard Burton reported seeing "the broad-leaved bread-fruit" at Lagos in 1862 and noted that although it had come originally "from the far Polynesian lands," in Africa it had "taken root like an indigen." A 1948 compendium of the "useful plants of West Tropical Africa" recorded some local terms for breadfruit including *tsho yĕlĕ*, meaning "yam borne on tree" in the Ga language. It seems, though, never to have reached a status beyond "one of the minor tree crops" of the region, as assessed in 1992 by the historian Stanley Alpern, who reminded his readers that "in introducing crops [into coastal West Africa and its offshore islands], Europeans were not being altruistic." Rather, the most important use of the crops they planted was "to feed the slaves brought from the African mainland" and awaiting transport to the Americas. With this history, it's no wonder breadfruit attained a reputation, an association with slavery, similar to that which it developed in the Caribbean.[1]

But Samuel thought that might change and he had two reasons to be hopeful. First, breadfruit's closely related ancestral crop, breadnut, had been introduced successfully even earlier and had been creolized into both Ghana's landscape and cuisine. In 1843 the Basel Mission—a Swiss Christian missionary society—convinced twenty-four Jamaican Moravians to immigrate to Akropong, a town thirty miles (about 50 km) inland from Accra, "to show the natives that there are christian negroes who cultivate lands." According to Carl Christian Reindorf, a "native pastor of the Basel Mission," these Jamaicans brought breadnut, which, according to Samuel, Ghanaians now consider a native crop. Could the same acceptance be in breadfruit's future? It may be challenging. In the Volta region, Samuel recalled seeing a group of children kicking around a fallen breadfruit like a football. "They didn't know the value," he explained.[2]

The second reason for Samuel's cautious hope was his faith. I asked him why he persevered with breadfruit after the discouraging advice he had received from the agricultural researchers. "Because God spoke to me," he simply replied. Throughout our meeting, during which we ate breadfruit cooked three different ways and drank honey-sweetened tea steeped from dried breadfruit leaves, Samuel spoke of breadfruit with the passionate zeal of a missionary sharing the Gospel. "There is nothing bad about this tree. Everything is useful," he preached.

After his vision, it took Samuel two years to obtain his first three breadfruit saplings. He planted two on a plot of land that he farmed and one in the courtyard of his church in Accra. This churchyard tree, planted in 2012, stood more than thirty feet (about 10 m) tall, and its limbs were heavy with fruit when I met Samuel in 2023. The tree shaded dozens of potted saplings—all taken as root shoots from this and other trees Samuel has acquired over the years. He now propagates breadfruit trees and distributes them to the members of his congregation, "to some ten, to others five," he explained, an allusion to Jesus's parable of the talents.[3]

I had come to Ghana to learn about breadfruit's third major journey, which is currently underway. Carrying breadfruit along on its first journey, beginning thousands of years ago, ancestors of today's Pacific Islanders left the intervisible islands east of New Guinea and ventured out into the Pacific, fueled by breadfruit, taro, and the confidence that their navigational skills could deliver them to new islands and show them the way back again. Breadfruit's second major journey, taking place about 250 years ago, provided sustenance for another human migration: the large-scale forced relocation of enslaved Africans to the Americas. Journeying across the oceans from Tahiti, breadfruit found a new home in the islands of the Caribbean, where it has now been creolized.

Of course, the Pacific voyages of discovery and the Atlantic slave trade were very different migrations, one of heroic exploration and the other of cruel enslavement. But in both histories, breadfruit provided vital sustenance to people as they attempted to etch out a living in their new island homes, whether those islands were chosen or not. The Pacific Islanders planted breadfruit wherever they settled. The enslaved Africans in the Caribbean survived, in some cases barely so, on breadfruit and other provisions introduced there as their foods. Many experts believe that the potential of this crop to provide for people's nutritional needs has not yet been achieved on a global scale. Breadfruit's third major journey, which includes spreading out and traveling to remote, food-insecure regions of the tropical world, is intended to realize that potential.

About a year before traveling to Ghana I went to Chicago to learn about the use of breadfruit in the fight against hunger and other sustainable development work. Just north of the city, the wealthy lakeshore community of Winnetka is home to a nonprofit organization called the Trees That Feed Foundation, whose mission involves planting breadfruit trees wherever they might feed hungry people.

The founders of Trees That Feed, Mike and Mary McLaughlin, an ebullient retired Jamaican immigrant couple, welcomed me into their home to discuss their organization's contributions toward sustainable development. In 2015 the United Nations approved a plan titled "Transforming Our World," which introduced seventeen Sustainable Development Goals, ambitiously intended to be achieved by 2030. The goals were eventually given succinct, catchy titles like "No Poverty" (Goal 1) and "Climate Action" (Goal 13) with more in-depth explanations provided in the supporting documents. Together these seventeen goals serve as guideposts for Trees That Feed and other organizations working to improve the well-being of the earth and its people. The framework of the Sustainable Development Goals allows you to see the topical reach of your efforts quickly, by simply listing the goals that are related to the work you do.[4]

Mike and Mary are focused primarily on three specific Sustainable Development Goals: "No Poverty," "Zero Hunger," and "Climate Action." The first two are bound up together in breadfruit's role as a food-producing tree, captured pithily in the very name of the McLaughlins' organization: Trees That Feed. The last, "Climate Action," is related to breadfruit's absorption of atmospheric carbon through photosynthesis, like every other plant.

It's been said that the best time to plant a tree was thirty years ago and the second-best time is today. To this end, the Trees That Feed Foundation is not alone in its mission to plant breadfruit trees as a form of sustainable development. The Breadfruit Institute in Hawai'i, Maryland-based Trees for the Future, the Breadfruit Line organization in the United Kingdom, the Australian government-funded Pacific Agribusiness Research for Development Initiative, the Barbados-based Caribbean Youth Environmental Network, a Honduran organization called Support Roatán, and the Jungle Project in Costa Rica are busy planting breadfruit trees all around the world. Today, through the work of these organizations, breadfruit has been planted throughout tropical Africa, the Caribbean, Central America, the Pacific islands, and South and Southeast Asia.[5]

Breadfruit may be particularly suited to help achieve the Sustainable Development Goals, since it is considered to be a "neglected and underutilized" crop, meaning that it is not among the major staple crops of the world but has unrealized potential to contribute to global food security and sustainable development. Breadfruit is also one of thirty-five crops listed in Annex I of the International Treaty on Plant Genetic Resources for Food and Agriculture, a compilation meant to include only those species "most important for food security." The promotion of neglected and underutilized food crops can reduce the risk exposure that results from an overreliance on the very narrow base of crops—rice, wheat, corn, and potatoes—that account for 60 percent of the human energy supply.[6]

Arriving in Ghana, I met up with Kwesi Agwani, an extension officer with the Ministry of Food and Agriculture, based in Ghana's Central Region and associated with the University of Cape Coast. Kwesi lives in Elmina and holds a PhD in agriculture and development studies. He combines a scholar's expertise with a local's rapport as he makes his rounds, visiting farmers throughout his district, offering advice and support. My plan was to ride along with Kwesi for several days as he checked in on farmers who had previously received donated breadfruit saplings to see how their trees were doing.

In Kwesi's large white 4 × 4, joined by a few members of his staff, we traveled the rutted red dirt roads of the Central Region, passing termite hills taller than any of us stood. When the roads inevitably narrowed into tracks, we would park the vehicle and hike into the remote farms where the breadfruit trees were planted. Farmers accompanied us to lead the way and to receive advice from the government experts. Twelve years previously, with funding from the Breadfruit Line, Kwesi had arranged to import two hundred breadfruit saplings, all propagated at a tissue culture lab in Germany, for distribution to farmers in his district. Those trees were all now at various stages of growth—some producing fruit abundantly, others showing signs of neglect. Curiously, those trees planted closest to the farmers' homes nearly always bore more fruit and appeared healthier than their field-planted counterparts. The extension agents and the farmers discussed this phenomenon and decided that the trees' roots must be finding water sources near houses—perhaps leaking pipes—that allowed them to grow and produce so well. This made sense to me, but I also remembered the advice I had heard from Hinano Teavai-Murphy, who regularly implied sentience among breadfruit trees. As we sat on a lanai looking out over Cook's Bay on the island of Mo'orea, Hinano had told me, "If you want the breadfruit tree to produce a lot, plant it by your house. That way he knows you're there to take care of him." Some people talk to their plants; others sing to them. Sometimes, though, simply keeping your trees close by is enough to maintain the relationship necessary for bountiful production.

Figure 11.1 Daniel Conduah harvesting breadfruit near Elmina in Ghana. (Photograph by Russell Fielding.)

For breadfruit to be an effective tool for sustainable development more is needed than just the distribution and planting of saplings. There must be a market for the fruits and other products of the trees. To this end, Kwesi and his Ministry of Food and Agriculture colleague, Victoria Abankwa, have been promoting breadfruit at regional farmers markets and food fairs. At an outdoor beachside café after a long day of fieldwork, with a local Afropop band playing in the background, Victoria told me about a recent experience. She had brought several dozen breadfruits to sell at a local farmers market and, since most potential customers would be unfamiliar with breadfruit, she planned to cut one up and cook it onsite so she could offer free samples. She had set the price at the very reasonable five Ghanaian cedis per breadfruit (about US$.40), but when people at the market tasted Victoria's fried breadfruit wedges, demand surged. She quickly raised the price to eight cedis each and still sold out within minutes.

I got to watch Victoria's culinary marketing firsthand when we returned from a long day of farm visits with several just-picked breadfruits in the back of Kwesi's vehicle. We arrived at the sprawling government compound where Victoria's office is located to find Solomon Ebow Appiah, the municipal chief executive, whom Kwesi referred to as "the mayor," standing on the building's front steps to greet us. As Kwesi and I made small talk with Mayor Appiah, Victoria hurried off to prepare a breadfruit for him to taste. A small basket of breadfruit wedges, fried in palm oil to a perfect golden crisp and lightly seasoned, soon made its way to the mayor's office, where he was able to get his first taste of this novel crop that had generated such a buzz. "Delicious!" he declared, underscoring his support for Kwesi's breadfruit-planting mission.

The main sustainable-development benefit of breadfruit is food production. Every breadfruit tree planted in a food-insecure region of Ghana—or any other country—and nurtured to produce fruit represents a step toward achieving the United Nations' Sustainable Development Goal Number 2: "Zero Hunger." But a productive breadfruit tree will do more than just feed the people who care for it: it will also provide income. Like other breadfruit-planting organizations, the Trees That Feed Foundation works with small-scale breadfruit growers worldwide to create local markets for breadfruit and to build capacity to make value-added products from it, like a scaled-up version of Victoria's free samples at the Ghanaian farmers market.

The process goes like this. If a country does not already propagate breadfruit trees, Trees That Feed will oversee and pay for the importation of hundreds of breadfruit saplings, usually from a tissue culture lab in North America or Europe. In places where propagation efforts are already in place, Trees That Feed will purchase locally grown saplings. The saplings are destined for a local plant nursery, a place where they can be tended by experts at keeping plants alive and healthy. Potential growers then fill out a tree request form on the Trees That Feed Foundation's website and Mike and Mary personally evaluate their requests. During my visit I watched as Mary pored over a recent request from Uganda, trying to ascertain whether the applicant would be a responsible grower. If approved, the applicant receives a digital coupon that can be exchanged for a set number of saplings at the nursery. Once the exchange is made, Trees That Feed pays the nursery for the saplings and the grower takes them home to plant. Mary told me that she only asks two things of her breadfruit recipients: keep the trees alive and put them on a map. The foundation keeps an updated map not just of the countries where they've supported breadfruit planting but of the GPS coordinates of each breadfruit tree planted through their work. New pins are always being added to the map, which is beginning to fill in throughout the world's tropical regions.

After receiving the trees, some growers don't need any additional help. These expert farmers are ready to add breadfruit to their thriving agro-businesses as soon as the trees begin producing fruit. Others, though, can benefit from guidance to help make their breadfruit production profitable. To this end, Trees That Feed offers both business and technological support in pursuit of the UN's Sustainable Development Goal Number 1: "No Poverty." On the business side, as protection against risk, in certain places the Trees That Feed Foundation offers to buy all the food products that growers or producers want to sell. In Haiti, for example, some enterprising young bakers have developed a recipe for breadfruit-based *konparet*, a dense, sweetened, sconelike bread made with coconut and ginger. Bakers who make *konparets* are free to sell them to whomever they wish, but Trees That Feed makes the endeavor less risky by offering to buy any surplus, which the foundation usually then donates to local schools for student lunches. With regard to technology, Trees That Feed often donates equipment such as solar dehydrators and flour mills to help new breadfruit-based businesses get started.

Finally, planting breadfruit trees makes progress toward Sustainable Development Goal Number 13: "Climate Action." As the planet's climate changes, trending toward warmer average global temperatures, the world's temperate regions are undergoing a process scientists call *tropicalization*. This means that animal and plant species normally seen in the tropics are finding suitable habitats in new areas, at latitudes further from the equator than those they could tolerate before. Some domesticated crops already established in temperate zones are becoming difficult to maintain as the climate warms. For example, one recent study predicted that the wine regions of southern Europe are likely to be negatively affected by warming temperatures, meaning both lower quality and quantity of grapes. The study came with a silver lining, though: "high-quality areas for viticulture will significantly extend northward in western and central Europe." So vineyards may become untenable in southern Europe even as new wine regions emerge in the north. How about a nice Scottish red or a rosé from Sweden?[7]

Tropical crops like breadfruit are also experiencing range shifts. Two recent studies have made predictions about breadfruit's future geography under predicted climate change scenarios. Both found that the overall area where breadfruit can grow would expand significantly, with little to no loss of existing range since breadfruit is tolerant of extremely high temperatures. One study referred to breadfruit as a key component of "climate-resilient low latitude food systems," a claim that was picked up and run with by the popular media, spinning off articles that referred to breadfruit as a "climate change-proof food of the future."[8]

But breadfruit trees do more than simply tolerate climate change: they actively work to diminish it through carbon sequestration. Don't all plants capture carbon? They do, but not equally. In Hawai'i, environmental scientist Noa Kekuewa Lincoln and his graduate student, Chad Livingston, measured the carbon sequestration potential of breadfruit. This is done by determining the volume of the entire tree and the carbon density of its wood and other tissues. In older studies, this involved cutting down a tree and literally weighing all its wood. Lincoln and Livingston left their trees standing and opted for the less invasive method of measuring the tree's diameter at every meter along its entire trunk and every branch. They also weighed leaves and extrapolated out to each tree's entire foliage. From those measurements, it's possible to estimate the amount of carbon sequestered and correlate that figure with a value you can measure more simply: the diameter of a tree trunk at "breast height," standardized at 1.3 meters (4.25 ft) within the field of forestry.[9]

Lincoln and Livingston's results showed that breadfruit is good, but not particularly great, at sequestering carbon when compared to other tree species found in similar climates. They expected such a result, considering the traditional uses of breadfruit wood in Pacific boatbuilding and surfboard-shaping. Its lightness, which translates to buoyancy, was prized in these applications, but less dense wood also means less carbon packed into the grain. Lincoln and Livingston pointed out, though, that it's not necessarily appropriate just to compare breadfruit to other tree species. Breadfruit is a domesticated food crop. It's farmed. Efforts to promote breadfruit consumption and cultivation in Hawai'i and elsewhere in the tropical world often focus on its potential to replace common, imported starches like potatoes and rice. The carbon absorption power of breadfruit far surpasses these crops, which are usually grown as annuals, meaning they're plowed up and replanted every year and are transported to most places where they're consumed—the carbon emissions of the fossil-fuel-powered ships, tractors, trains, and trucks far surpassing the carbon sequestered by the plants bearing the crops. By contrast, a breadfruit tree may be left to grow and absorb carbon season after season, all while producing a locally consumable food, for decades or more.[10]

Referring to coal, natural gas, and petroleum collectively as "fossil fuels" has the unfortunate effect of creating a popular association between oil and dinosaurs. The Sinclair Oil Company's well-known apatosaurus mascot, "Dino," endearing as it is, probably adds to the confusion. But fossil fuels are only tangentially related to what we normally think of as fossils. Coal, petroleum, and natural gas do not come from dinosaurs. Instead, the term refers to the subterranean and nonrenewable aspects of these commodities. In this way, fossil fuels are more closely related to fossil waters, underground aquifers that were filled long ago and aren't being recharged by percolating rainfall anymore, either due to changing weather patterns or continental drift. People must be especially careful not to rely too heavily on fossil resources—be they fuels or waters—since by definition they aren't being replenished, at least not at rates comparable to their use.[11]

Consider coal, which is made up of the decayed, compressed, and carbonized plants that grew in swamps long ago. Yes, there are still swamps, and the processes that created the coal we burn now are still ongoing so, technically, coal deposits are still being formed today. But the rate of coal formation is pitifully slower than the rate at which we burn it. The history of the world has seen far more years of coal formation than it has of coal burning.

When we burn fossil fuels today, we're using solar energy stored from sunny days long ago. Combustion requires oxygen, which is available in the atmosphere and combines with carbon and hydrogen released from burning fossil fuels to make CO_2 and water. This process produces heat, which power stations use for energy, or motion, which pumps the pistons in our vehicles' engines. The carbon in the gasoline in your car's tank was broken off from molecules of CO_2 that floated in the atmosphere before humans even existed. Photosynthesis performed by some prehistoric plant captured carbon, integrated it into stems, roots, or leaves thriving in a swamp through which long-extinct creatures splashed and slithered. When that plant died, it decomposed in the anoxic waters of the swamp and, with time and pressure, became the lump of coal burning right now to power the lamp by which you read. Biologist Jeffrey Dukes has calculated that every year humans burn more than four hundred years' worth of this "buried sunshine." To my thinking, this is a clear case of unsustainability.[12]

Classicist Chris McDonough has a spin on Dukes's concept, though: fossil fuels aren't just buried sunshine; they're also buried sins. In a 2019 essay, McDonough recounted a story he heard from a woman at a local park, which tracked the Genesis account of Noah's flood quite well in plot, if not in keeping with the King James language. The woman's story began with God having just flooded the earth in response to mankind "killing each other and raping each other, and being evil just however they could." The dinosaurs, the other animals, and the plants—*even the plants!*—were just as guilty as humans, she said, and they too got what they deserved. Noah and his family were spared and went on to repopulate the planet. Millennia later, one of Noah's descendants—we're all Noah's descendants, of course—named Rockefeller started drilling for oil and "getting richer than anybody oughta be." The woman paused. McDonough sensed where she was going.[13]

"You know what oil is, don't you?" she asked. Oil is made of everything that lived on earth long ago, she explained, and when we drill for it, we bring to the surface "all that evil that God buried under the mud." We put this "primordial evil" into our cars, houses, airplanes—everywhere. "We're taking all the stuff God hates so much that he had to kill it in the worst and most final way he could think of," and it's coming back to haunt us. "You can't pump all that evil up out of the ground and not expect it to make everything else just as evil." It's as if all those evil people—and dinosaurs and animals and plants—are getting a second chance to destroy the world. And this time, we're abetting them.[14]

When CO_2 enters the atmosphere, whether through the combustion of carbonized plant remains or the release of antediluvian demons, it contributes to the greenhouse effect along with other gases like methane and water vapor. Greenhouse gases trap the sun's heat and keep it within the atmosphere, where it continues to heat the planet even as more of the sun's energy arrives every day. Breadfruit, like other plants, has the power to absolve our climatic sins. Children understand that plants breathe in carbon dioxide and breathe out oxygen. This is true, if you allow for a generous definition of the word "breathe." A more complicated, but not necessarily more correct understanding of photosynthesis is that plants use sunlight to catalyze a chemical reaction that breaks apart carbon dioxide molecules in the atmosphere, discards the oxygen as a byproduct, and combines the carbon atoms with water to make a sugar called glucose, otherwise known as plant food. Removing CO_2 from the air gives the solar energy reflected from the earth's surface a better chance at escaping the atmosphere so its time of heating the planet can come to an end. This process is quite literally the opposite of fossil fuel combustion: instead of emitting CO_2 into the atmosphere, plants absorb it and break it down into carbon and oxygen, keeping the carbon for themselves and releasing the oxygen for us to breathe. And it's but one of the many services the earth and all its magnificent systems provide for us, every single day, for free.[15]

In 1997 a team of researchers led by ecological economist Robert Costanza set out to determine the monetary value of the earth's natural capital, what the United Nations has called the "multitude of benefits that nature provides to society." Consider how difficult, how expensive, it would be to make all that oxygen ourselves if every plant on earth were to disappear. We would, presumably, have to use fossil-fuel energy to desalinate and electrolyze ocean water, breaking down the H_2O into hydrogen and oxygen. Thinking of doing this at the scale required to match the amount of oxygen produced through photosynthesis brings the value of the earth's natural capital into focus. And it's not just oxygen. The biologic, hydrologic, and geologic processes producing soil as they have since before the first human ever planted the first seed to start the agricultural revolution, the processes of pollination and water purification necessary for us to eat and drink: all happen freely and naturally in functioning ecosystems. Placing a price tag on all of that, it's a fool's errand. Even Costanza and his fellow experts who calculated the value acknowledged the "many conceptual and empirical problems" with what they set out to do. But still, they settled on a number. That number was US$33,000,000,000,000. Thirty-three trillion dollars.[16]

By comparison, the global gross national product in 1997, when Costanza's team published their paper, was approximately $18 trillion, meaning that the earth's natural capital was just about double the value of all human capital combined. But, just like anything else, the price of all these ecosystem services is subject to change. In 2014, Costanza and his colleagues updated their numbers, finding that the total value of the earth's natural capital had risen to $125 trillion. Inflation? Partially, but the higher price is also due to scarcity: the principle behind the rule of supply and demand. As the human population continues to grow, placing greater demand on the earth's ecosystem services, we lose between $4.3 trillion and $20 trillion worth of natural capital per year. What kind of land-use changes cost that much, and why would anyone do anything so costly? The key to understanding this loss lies in the notion of scale.[17]

When it comes to money and time, most people find it difficult to think big. Exceptionally large numbers can become meaningless to us. Be honest: do you really have a clear concept of how much $125 trillion is? I don't. Millions, billions, and trillions are all members of the Very Big Number Club, and we can forget sometimes that they aren't all members in equal standing. It may help to imagine counting at the rate of a number per second. This is easy enough at first, but try reading out "seven hundred forty-nine thousand, eight hundred and twenty-three" in one second flat, then moving right on to "seven hundred forty-nine thousand, eight hundred and twenty-four," and so on. If you could keep up the pace without stopping, even to sleep, it would take you a little longer than eleven days to reach one million. What about one billion? At the same rate, again with no sleep or breaks of any kind, you'd be counting for almost thirty-two years. A trillion? The concept of numbers has barely existed long enough for counters to have reached a trillion, even if the first prehistoric tally-stick carver had begun at *one* and each generation took over the count when the previous one died.[18]

The value of the earth's natural capital, as calculated and recalculated by Costanza and his colleagues, is simply too great to comprehend, much less to pay. The pollinators who fertilize our crops, the phytoplankton and terrestrial plants who generate our oxygen, the lightning that fixes nitrogen in the soil, and on and on and on. These laborers do not rest, nor do they seek payment.

With this in mind, consider a breadfruit grove. A stand of trees silently providing fruit for food and wood for shelter; absorbing carbon and stabilizing soil; creating habitat for all manner of bats, birds, insects, and more. How much is that worth, multiplied out across all the acres under breadfruit girdling the earth's tropics? Impossible to calculate? Let's say you made an estimate anyway, like Costanza and his team did. If you were to approximate the value of the ecosystem services that breadfruit provides, and felt obliged to pay it back, to whom would you write the check? Kū, the Hawaiian deity said to have given his mortal family the first ʻulu tree through an act of self-sacrifice, doesn't need your money. But perhaps there is someone who does.

Economists have recognized the real value of natural systems for a very long time. In 1776, Adam Smith wrote in *The Wealth of Nations* that "nature labours along with man; and though her labour costs no expence, its produce has its value, as well as that of the most expensive workmen." The value of nature's produce, as Smith called it, is what has come to be called "natural capital," or, in Costanza's value calculations, "the multitude of benefits that nature provides to society." People's choices, particularly regarding how to manage land that's been placed under their control, can have profound impacts on the degree to which "nature labours along with man" and the kind of natural capital that's created.[19]

For example, a tropical farmer might consider either raising livestock, like cattle, on her land or planting a long-lived tree crop like breadfruit. Of course, it's possible to do both— Duane Lammers's cows graze happily under ʻulu trees at Hana Ranch on Maui—but let's assume our prospective farmer has only enough funds to invest in one commodity. With either choice, the farmer would produce a marketable product—milk or meat from the cows, breadfruit from the trees—that could be sold to generate income and a return on her investment. These two options, though, involve very different kinds and values of natural capital. The cattle produce methane, a greenhouse gas more powerful than CO_2, in abundance through their digestion, specifically their belching and flatulence, actively contributing to climate change. This is a big deal because there are so many cows in the world. A comparison popularized by tech billionaire and philanthropist Bill Gates in 2018 considers that "if [cattle] were a country, they would rank third in greenhouse gas emissions" after China and the United States. The breadfruit trees, by contrast, absorb CO_2 from the atmosphere, each one doing its small part to fight climate change. Compared to the Republic of Cattle, the Republic of Breadfruit, then, is not just carbon-neutral but carbon-negative.[20]

How should the farmer choose? Any economically minded entrepreneur would consider, among other factors, her startup costs and the time until a return on the investment could be realized. Cows mature quickly; in a year or two a newborn calf can be full-grown. Breadfruit trees, even when propagated by air-layering or tissue culture, usually take much longer to begin producing fruit. Of course, the trees aren't sitting idle during that time—they grow, taking in atmospheric CO_2, extending their root systems, stabilizing the soil, and spreading their leafy canopies, creating shade and habitat. These are all ecosystem services that remain unaccounted in the farmer's ledger book. Yet, all this time, a growing steer is just eating grass and passing gas.

During the late twentieth century, environmental economists began to suggest that society might consider actually paying land managers like farmers who make decisions that result in greater natural capital. A scheme now referred to as "payments for ecosystem services," emerged, with the goal of incentivizing decisions that might not yield as much profit, at least at first, as compared to alternative options, but that would benefit society through the creation and maintenance of natural capital. These payments are intended to nudge farmers, or other land managers, toward choices that might be harder to make based only upon the expected return on investment, but that would result in more, or better, ecosystem services for us all. Sometimes payments for ecosystem services come in the form of direct cash transfers; other times it's an in-kind payment, like the delivery of breadfruit saplings, free of charge, the costs having been covered by Trees That Feed or one of many other tree-planting charities.

One of the most famous examples of a successful payments-for-ecosystem-services plan involves the municipal water system in New York City. New Yorkers love their tap water. Bagel and pizza connoisseurs claim that it's the water used in the dough that makes these iconic foods taste so much better in New York than anywhere else. Or simply when made with New York water: bagel shops and pizzerias in states as far away as California have been known to import casks of New York City tap water for use in their kitchens, and, of course, to advertise their products as being more authentic for it.[21]

Unlike most major cities in the world, New York is able to avoid the cost of extensive water-treatment and filtration systems because of the city's investments in the ecosystem services provided within the watersheds from which it draws its water. By paying landowners to implement, or continue, sustainable farming and forest-management practices, New York's city government has saved billions of dollars that otherwise would have been spent on water treatment. Instead, the city pays the landowners whose fields, soils, and trees filter the water naturally. Could New York's success be replicated elsewhere?[22]

Back in Ghana, Kwesi stopped walking abruptly one day as we were leaving another farmer's breadfruit grove. He showed me a text message on his phone, but without context, all I could see were big numbers. "This man owns many hectares of land in the Northern Region," Kwesi explained. "He wants to plant it all in breadfruit." For the rest of the day, Kwesi's attention was given to this large landowner: How many breadfruit saplings did he want? How long would it take to produce that number in Ghana's recently established tissue culture lab? And what would this mean for Ghanaian food production?

I shared in Kwesi's enthusiasm, but, as we drove the dirt roads through the mud-brick villages on our way back to Elmina, I couldn't help but think of the potential growers, not just of breadfruit but of all kinds of crops, we were passing along the way. The lack of capital is, of course, a barrier to much business development, but when the enterprise that one would start, if only there was a way to cover the up-front costs, is something as deeply connected to the public good as tree planting, it just makes sense that society would want to help that person get started.

When I got home from West Africa I had lunch with Christopher Beyuo, a Catholic priest whose pastoral assignment in South Carolina was coming to an end and who would soon be returning to his home country of Ghana. Christopher spoke of his dedication to service, not just in the church but in the community. He was especially interested in food security and had heard about breadfruit. He wondered if this unusual tree crop might be a part of the solution to hunger in his home village of Ketuo in Ghana's

Upper West Region. I told him of Samuel Glory Abbey's vision. I described the work that Kwesi and Victoria were doing along the Cape Coast to see breadfruit trees thriving in the soil and breadfruit products being brought to market. But since visions such as Samuel's cannot be implemented without capital, and since you don't need an agricultural extension officer until you're in a position to plant, I also put Christopher in touch with the McLaughlins at Trees That Feed.

CHAPTER XII

Super Food

Breadfruit's third global journey involves more than just its role in sustainable development. Noted for its health benefits and culinary versatility, breadfruit is being discovered by gourmands and pursuers of wellness as a starch with better nutritional value, more sustainable production, and more interesting culinary potential than competitors like potatoes and rice. Breadfruit is beginning to appear as a novel ingredient on restaurant menus and in grocery store products throughout Europe and North America. In advertisements and news articles that hype its potential, breadfruit isn't just called a novel food, it's often called a "superfood."

There are no superfoods. The term, despite its ubiquity within popular nutrition-focused media, has no formal definition among experts in any scientific field. According to author and nutritionist Matt Fitzgerald, "science does not support the separation of foods into the categories of mere foods and superfoods." Still, many individual food products have been called superfoods and many others have been proposed as *potential* superfoods, as though there were some kind of vetting process. There is not a process, nor is there any governing body, meeting regularly to vote on which nominees are given "superfood" status and which are not. People—journalists, nutritionists, researchers—simply suggest that certain foods be considered superfoods, and those suggestions can be ignored or accepted by the public. In recent years the roster of proposed superfoods has grown to include such examples as açai, almonds, blue-green algae, blueberries, broccoli, cacao, chia seeds, cranberries, goji berries, kale, kava, kimchi, kombucha, kudzu, leafy greens, legumes, matcha, miso, moringa, olive oil, quinoa, salmon, seaweed, tomatoes, turmeric, and yogurt, among many others. In 2018, European researchers proposed that something called "cockroach milk" be considered a superfood, and I think we should end our list right there.[1]

Given the lax requirements for consideration in the category—the existence of which scientists don't even recognize—the concept of superfoods has attracted a fair amount of criticism. Some critics have suggested a compromise: accept the term in recognition of its popularity and establish rigorous criteria for attaining the status, such as requiring that a proposed superfood be associated with a lower incidence of certain diseases or that it contain a certain number of essential nutrients.[2]

The human body requires at least forty "essential nutrients" for life. This list comprises vitamins, minerals, amino acids, and fatty acids. No single, natural food item exists that includes all the necessary nutrients. Other animal species have evolved to become dietary specialists: koalas famously consume eucalyptus leaves almost exclusively and pandas are picky about bamboo. But we humans must vary what we eat. We have choices to make. Humans face an omnivore's dilemma, as author Michael Pollan explored in his book by that title. The dilemma involves making wise choices about what to eat in terms of health, cost, taste preferences, cultural relevance, justice, and sustainability. Some foods we might choose are clearly bad choices, owing to their unhealthy ingredients, unaffordability, or sourcing that can be unethical, unsustainable, or both. Others are much better: affordable, nutritious, and produced in fair and environmentally friendly ways. The best choices of all, those popularly considered to be superfoods, excel in several of these areas; they're especially efficient, providing the most bang for your buck, or benefit for your bite. And breadfruit is beginning to be counted among these.[3]

Alleged superfoods often provide antioxidants, chemical compounds that protect the body from the harmful effects of free radicals. Free radicals are atoms with unpaired electrons. Thinking back to basic chemistry class, you'll recall that an atom has multiple orbitals—the "levels" in which its electrons can go. Each orbital can accommodate two electrons. If we allow ourselves to anthropomorphize subatomic particles for a moment, electrons *want* to pair up; they don't like to be alone. This is why nitrogen is usually found as N_2, oxygen as O_2, and so on. Atoms join in what's called a covalent bond to share electrons so that none remains unpaired.[4]

When this pairing-and-sharing fails to happen, the atom becomes a free radical, on the hunt for electrons. When it finds them, the free radical reacts, grabbing electrons and setting off a chain reaction by which each atom that has had an electron "stolen" by the free radical now needs to take one from a neighboring atom. Some of the most harmful interactions with free radicals happen in the DNA and are associated with aging-related degenerative diseases like cancer, dementia, and heart disease.

Many free radicals are oxygen-based. They're called "oxidants" and their reaction with other substances in your body is called "oxidation." To counteract the potentially deadly effects of these hyperreactive oxidants, antioxidants rove throughout the body like bounty hunters, seeking free radicals. When an antioxidant finds a free radical, it offers up an electron without becoming unstable itself, thus neutralizing the free radical and preventing oxidation. This defense mechanism is good, but not perfect. Some free radicals still make it through to oxidize DNA, thus damaging the cell's genetic code and, potentially, causing life-threatening diseases or mutations.

Fortunately, repair enzymes are quick to respond to damaged DNA, and they succeed most of the time. To increase the body's success rate at surviving the oxidizing effects of free radicals, nutritionists recommend including in your diet plenty of fruits and vegetables, especially those that are high in antioxidants. Laboratory analyses of breadfruit have found levels of antioxidants comparable to those of apples and citrus fruits and far greater than potatoes and other starches likely to be replaced by breadfruit in one's diet.[5]

But it's not all about antioxidants. Breadfruit is a good source of energy with complex carbohydrates at levels comparable to other traditional starchy foods like potato, sweet potato, and rice, and with higher concentrations of some vitamins and minerals. As such an abundantly productive tree, breadfruit produces food containing these nutrients in quantities unmatched by nearly any other crop, when you consider the amount of land or agricultural effort given over to its cultivation. The multiple cultivars of breadfruit differ in their nutritional profiles and the timing of when the fruit is harvested, in terms of its ripeness, also affects its nutrition. In the Pacific islands, traditional methods of preparing and preserving breadfruit can serve to concentrate its nutrients, yielding a product with a longer shelf life and denser nutrition than fresh fruit. As we have seen, when dried and milled, breadfruit produces a flour that's gluten-free.[6]

We hear about gluten all the time. Restaurant menus often label some food items "GF" for gluten-free. Gluten is found in foods made from wheat and a few other grains like barley and rye. These grains naturally contain two proteins—gliadin and glutenin—that can combine to form an elastic substance called gluten. It's normally formed when wheat flour is mixed with water and kneaded to make dough. One of gluten's most noticeable contributions to food is texture. Bread dough gains its stretchiness from gluten, bagels and baguettes attain their characteristic chewiness from gluten, and any yeast-risen dough gets its spongy texture because gluten traps carbon dioxide bubbles, emitted as the yeasts feed on sugars in the dough. More kneading means more gluten, which is why certain recipes advise you not to overmix your dough. Biscuits, for example—not the crunchy British cookies but the bread made famous in the American South—should be light, flaky, and fluffy, not chewy. Other breads, like pizza crust, should have plenty of gluten, and these recipes often call for more kneading.[7]

Most people want gluten in their food because it imparts the textures we like. For the 1–3 percent of the world's population with celiac disease, an autoimmune disorder that causes damage to the small intestine when gluten is consumed, however, avoiding gluten is essential, a matter of life and death in extreme cases. Still more people without a celiac diagnosis report feeling better when they've eliminated gluten from their diets. The problem is that gluten is so ubiquitous. The Mayo Clinic's website on celiac disease lists twenty-three food groups that one should avoid unless they're explicitly labeled "gluten-free." Some of the categories are quite broad: beer, bread, cereals, pastas, and salad dressings all appear under the heading "Avoid."[8]

Because gluten is found in so many foods, maintaining a familiar diet without it can be difficult. You could simply give up breads, cereals, pastas, and more, but many people prefer instead to find alternative foods that simply replace the offending ingredients. When ground into a flour, breadfruit can serve as a wheat substitute, replacing a potentially harmful food in the diets of people with celiac disease or gluten sensitivity. But so can lots of other plants. Flours are made from almonds, cassava, chickpeas, corn, oats, potatoes, quinoa, rice, and many other starches and grains. Taste-testers rate some of the products made from these alternative flours higher than others. A 2020 article in *Outside* magazine's online nutrition series advised that "You can grind almost anything into a flour, but that doesn't mean you should."[9]

Helpfully, several food-science scholars, professional and amateur, have taken on the responsibility of comparing the various wheat flour alternatives. Reviews of breadfruit flour often remark upon its neutral-to-nutty flavor, its high concentration of nutrients, and the abundance and sustainability of its production. On most of these counts, breadfruit flour easily surpasses its competitors.

Breadfruit contains vitamins, minerals, and carbohydrates. It supplies antioxidants and can be made into a gluten-free flour. It grows abundantly and sustainably. So, to those open to the idea that we should have a category called "superfoods," does it qualify as one? According to popular news sources ranging from the BBC to *National Geographic*, it does. *Time* magazine even called breadfruit a "REAL Superfood," as opposed to the alleged imposters named in the article like kale. In 1967, physician Derrick Jelliffe introduced the concept of "cultural superfoods," which he defined as the dominant staple of a society, foods which "because of their importance for the survival of the particular community . . . often have semidivine status, being interwoven into local religion, mythology, and history."[10]

The relevance of Jelliffe's concept to breadfruit becomes clear when we consider the entire tree, and its role in the cultures of the Pacific and the Caribbean, not just the nutritional value of the fruit. Think of the leaves producing oxygen while absorbing atmospheric carbon and sequestering it in the branches, trunk, and roots. Think of those roots holding back soil from eroding, the leaves and branches providing shade for humans and habitat for other species. Imagine an old grove of breadfruit trees fifty years from now, still providing these ecosystem services along with a regular supply of nutritious food, a subject for painters and poets, and, eventually, when a hurricane or a chainsaw determines the date of the trees' felling, wood for sculptors and builders of boats, homes, surfboards, and coffins. Breadfruit trees are part of the landscape, part of the culture, connecting people to their history and their ancestors, creolizing today in new soils and being revitalized on the Caribbean and Pacific islands where they had been neglected. Considering its ecosystem services and cultural importance, along with the bountiful production of nutritious fruit, even the staunchest critic of popular yet imprecise nutritional terminology can see why, as part of the cultural, natural, and nutritional landscapes where it grows and is eaten, some might consider breadfruit a superfood.

Breadfruit's emerging popular reputation suggests the need to take cautionary advice from the past. At the turn of the twenty-first century, quinoa—a grain produced by a plant native to the Andean highlands of Bolivia and Peru—was touted as the next big thing, the new superfood. Agricultural scientist Sven-Erik Jacobsen wrote in 2003 of quinoa's "significant worldwide potential," employing language similar to that with which breadfruit is discussed today. Both crops have received praise for their potential to contribute to "the world's most pressing problems, including climate change, drought, poverty, and malnutrition," as summarized by anthropologist Fabiana Li. Quinoa was grown traditionally by Indigenous farmers, but following the Spanish conquest of the Incan empire its cultivation began to decline and it was replaced in local diets by other grains, including the rice that the Spanish introduced. When quinoa arrived on the culinary scenes of Europe and North America, demand surged, leading to both positive and negative outcomes in the Andean communities where it grows. Sociologist Marygold Walsh-Dilley has highlighted the economic, environmental, and social conflicts that arose in South America following quinoa's rise abroad. While the definitions of "superfoods" are vague and schismatic, the trajectory that breadfruit has begun does seem to resemble the path quinoa took several decades before. It should be important, then, to ensure that breadfruit's rise as a "superfood" is done right.[11]

But for now, outside the Pacific and the Caribbean, breadfruit is still a niche crop. Why? One reason is that it tastes quite bland when not prepared well. In 2011, the *Wall Street Journal* proclaimed that breadfruit could be "the food of the future" except for one flaw: "it's all but inedible." This was hardly a novel opinion. The nineteenth-century naturalist Alfred Russel Wallace ate breadfruit regularly while conducting research in Southeast Asia. He was not impressed, writing that "breadfruit, prepared after the manner of slicing and frying . . . resembled fried slices of a friable cardboard." Even Diane Ragone, director emerita of the Breadfruit Institute, is said to have hated breadfruit the first time she tried it.[12]

But another way to think of breadfruit's blandness is in terms of its potential. Interviewed for an article in *Smithsonian* magazine, one chef called breadfruit "the perfect canvas for a culinary artist." Similarly, food historian Rachel Laudan has written that "very few plants are intrinsically delicious," explaining that "they take breeding, processing, and cooking to become delicious." Turning her attention to breadfruit's inherent blandness, Laudan advised that "if you want established delicious breadfruit dishes, then go to the South Pacific, the Caribbean, Southeast Asia or South India where cooks have long been serious about breadfruit." Culinary innovators throughout these regions, like those I met, are busy coming up with creative new breadfruit-based recipes to challenge its bland reputation. Their creations are anything but inedible. Even Alfred Russel Wallace was eventually convinced that breadfruit could be made delicious, later writing that "with meat and gravy it is a vegetable superior to anything I know either in temperate or tropical countries."[13]

What if you don't live in the tropics but still want to eat breadfruit? Outside of areas where it grows, breadfruit is still rather obscure. Despite the year-round, globalized bounty of tropical fruits available for purchase at nearly any North American or European grocery store, breadfruit is difficult to find except in cities with large Caribbean or Pacific Islander diaspora communities.

Part of the reason is its short shelf life. After harvest, growers must hurry to get the fruit to market before it begins to deteriorate. But other perishable fruits have short postharvest shelf lives, including avocados, grapes, mangoes, and strawberries. In these cases and more, careful handling and packaging, quick shipping methods, and the control of both temperature and moisture make it possible for consumers to buy fresh fruits in shops far from their places of origin. Breadfruit might be particularly challenging, both due to its perishability and the remote locations where it grows, but the techniques are available, if costly.

Another reason for breadfruit's absence from grocery shelves outside the tropics might be a simple lack of familiarity among both buyers and sellers. Journalist Amanda Fiegl recounted searching for breadfruit among the ethnic food markets of Washington, DC, in 2009 and being met with a confused grocer's question, "You want fruit or bread?" When Texas-based historian James McWilliams went on a similar quest in Austin a few years later, the produce clerk at his local Whole Foods Market simply responded, "Whoa!"[14]

I tried a similar search after a colleague told me that he had seen fresh breadfruits for sale at an upscale grocery store in New York City for $30 apiece. I called the Whole Foods Market in New York's Bowery neighborhood and asked for the produce manager. The voice on the phone sounded confused when I asked if breadfruit was in stock. "What are you looking for?" she asked, "a fruit whose name is *bread*?" In the background I could hear the manager conferring with another employee in Spanish, so

I tried a new approach: "*Pana. Estoy buscando pana. ¿Usted la tiene?*" I asked. "*¡Ay, pana!*" she said, sounding relieved that we had broken through the confusion. It was January when I called, and the produce manager explained that they sell *pana* when it's in season—usually late spring and early summer. I asked about the thirty-dollar price tag. It didn't seem unreasonable. "The price really varies depending on the supplier," she told me, "It could be lower or higher."

European and North American grocery stores may be slow to stock breadfruit, but online produce markets have stepped in to fill the niche. One of the best-established is a company called Fresh Direct, which delivers breadfruit, among lots of other produce, anywhere within New York City and parts of Connecticut and New Jersey. I don't live in that area, but, using the New York Botanical Garden's address in the Bronx for a hypothetical delivery, I was able to create an order for thirteen breadfruits—which reached the minimum price for same-day delivery—for just over $60.

Other North American cities support similar services, some with wider delivery networks. Miami has Vega Produce, which would ship thirteen breadfruits to my home in South Carolina for a little more than $100. Sasoun Produce, based in Los Angeles, ships breadfruit to every state in the country except Hawai'i, which makes sense. Through their website I was able to order a "Breadfruit Box," containing between eight and nine pounds of breadfruit, for $99. Dam Foods, located just outside Toronto, will deliver a single fresh breadfruit along with one roasted breadfruit—their specialty—to the Toronto Botanical Garden, whose address I used since Dam delivers only within Canada, for C$30 (about US$22).

London has no shortage of options for ordering breadfruit online, all with highly variable prices. Windrush Bay will sell a single breadfruit, to be delivered anywhere within the United Kingdom, for £10 (about US$13), plus delivery, but was sold out when I attempted to place a fictitious order. I was able to initiate an order with UK-based Exotic Express ("The Online Exotic Supermarket") for a single breadfruit—grown in Mauritius, the website specified—to be delivered to the Royal Botanic Gardens, Kew, for £32 (about US$41). A single breadfruit delivered by London Grocery would cost £63. That's almost $80 for one breadfruit! If you live in London and were considering ordering six or more breadfruits from London Grocery, it might cost less to fly to Jamaica and pick your own. So much for Rousseau's claim that "the fruits of the earth belong to us all."[15]

Because there is so much competition in the online tropical fruit delivery market, some companies have sought to distinguish themselves by creating a brand identity. The undisputed leader in this approach is Miami Fruit. Through its colorful social media accounts, Miami Fruit directs its nearly one million followers to a website where, depending on the season, you can order an eight- to ten-pound box of breadfruit to be delivered to your door for $147, plus shipping. In Miami, I caught up with Edelle Schlegel, cofounder of Miami Fruit, to learn more about her company's story.

Originally from California, Edelle went to Hawai'i to attend university and, she explained, "just to live in the tropics." There she met her partner, Rane Roatta, a vacationing Miamian who had started a fledgling bicycle-based fruit-delivery business back in his hometown. Rane must have made an impression because Edelle soon moved to Miami to help develop the business, mainly by marketing online through social media.

"It's just me with my phone taking a picture with fruit," Edelle explained. That approach may sound simple, but it worked. Miami Fruit's social media accounts feature fun pictures and videos, some close-ups of tropical fruits sliced open to reveal their edible interiors, others showing sunny scenes of Edelle, Rane, and their friends picking, slicing, and eating various fruits. These posts attracted followers by the thousands, a customer base Edelle attributes, at least in part, to her casual yet enthusiastic approach to marketing. "People online," she explained, "they know what's an ad and what's just somebody sharing what they're passionate about. So there's an advantage to not looking like an ad." Scrolling through Miami Fruit's social media posts certainly doesn't feel like watching advertisements. You're simply looking at images of young people, outdoors in the sun, happy to be surrounded by colorful and unusual fruits.

Most of Miami Fruit's clientele is online, not local to South Florida. Selling online opened Miami Fruit to a much wider and fruit-deprived clientele. "People who live up north just don't have access to a lot of this stuff," Edelle told me. As Miami Fruit grew, Edelle and Rane were able to buy land outside of Homestead and began to grow much of the fruit they sell themselves. The land they bought already had several breadfruit trees growing on it, which supplied the first few "breadfruit boxes" they sold.

Breadfruit, though, hasn't been their best seller. By far, most customers' preference has been the "variety box," which contains bananas—always bananas—plus whichever tropical fruits are currently in season. Prices range from $100 to more than $400, depending on the size and shipping method.

I asked why breadfruit hasn't been more popular. Edelle explained that it's likely due to its lingering unfamiliarity throughout much of North America.

"We're realizing now that we maybe got a little ahead of ourselves with some of the fruit that we offer," she admitted. "We could do a better job educating people about some of the unique stuff." Edelle then told me the story of Miami Fruit's foray into the durian market. "We were really excited about durian, which is the smelly fruit," she explained. "We sourced it from Malaysia, which is known for having the best quality and the most variety and we were very excited to start offering it in the United States. But we're realizing now that, uh-oh, a lot of people don't like durian here." Their response has not been to stop selling durian. Rather, Miami Fruit launched an educational initiative to teach people about unfamiliar fruits and, in doing so, to create demand. "We have to share more information and educate more people and also maybe figure out what people like, not only the rare stuff that we're excited about," Edelle said.

Breadfruit seems to fit, currently, within that category: a rare fruit that those in the know are excited about but that's still unfamiliar to many potential consumers in North America and Europe. It's likely due to this current obscurity that breadfruit isn't yet widely available at American and European grocery stores and can be obtained reliably, throughout much of North America, only through specialty online vendors such as Miami Fruit.

If the principle of supply and demand holds true, however, breadfruit's availability might soon increase. In 2023 both *Forbes* magazine and the *New York Times* carried articles repeating a prediction from Carbonate, a San Francisco–based marketing firm, that breadfruit would soon emerge as a food trend in the United States. Carbonate's report noted that breadfruit "is often described using superfood words," that it is "hearty, healthy, and sustainable" and that it's expected to be "widely embraced both in packages and on plates" in the near future.[16]

I called Leith Steel, senior strategist and head of insights at Carbonate, to learn more about what she saw in breadfruit that made it a likely trend. "Breadfruit's hitting a lot of buzzwords," Leith explained, "words like *health* and *sustainable*." She didn't see breadfruit's blandness as a problem and thought that chips were likely to be many people's first breadfruit experience. "It's a great alternative to a potato chip," according to Leith. They just look delicious and "people eat with their eyes first"—a truism in the food-marketing business. She did have reservations, however, about the name. "Bread. Fruit," Leith intoned slowly. "It's confusing and the name is horrible." We talked about William Dampier and the original English naming of breadfruit. Maybe it had more appeal during the seventeenth century than it does today. Now, it seems, people don't want their fruit to resemble bread. "I would not be surprised if breadfruit gets renamed," Leith predicted, and mentioned thinking that the Hawaiian word ʻulu sure is fun to say.

"There's always a hot superfood," Leith said, predicting that breadfruit is on the cusp of taking that title. Her track record with Carbonate is good. The trend forecast for 2021 rightly predicted that jackfruit would rise in popularity and the following year, Carbonate singled out Caribbean cuisine as an increasingly popular culinary style. Carbonate's forecast that breadfruit is on the cusp of broad popularity has its predecessors. As far back as 1928, an agricultural treatise on tropical crops advised that "slightly toasted slices" of breadfruit, "whole, as chips, or broken in small pieces are delicious eaten with hot butter or cream or in milk," and predicted, matter-of-factly, that these predecessors of today's breadfruit crisps "will some day be very popular."[17]

Not all innovative uses of breadfruit involve food. Perhaps inevitably, people have also found ways to turn breadfruit into alcoholic beverages. Author Amy Stewart wrote in the preface (which she calls the "Aperitif") to her book *The Drunken Botanist* about how "around the world, it seems, there is not a tree or shrub or delicate wildflower that has not been harvested, brewed, and bottled." True to this claim, breweries in Hawaiʻi, Sāmoa, and Tahiti have all produced breadfruit-based beers. Vintners in the Caribbean have produced several varieties of breadfruit wine. Distillation allows the creation of even more potent potables made from fermented breadfruit. Since 2018, the Royal Hawaiʻi Spirits Distillery in Honolulu has produced brandies, vodkas, and whiskeys with fermented and distilled breadfruit. Another breadfruit beverage distillery, based in the Caribbean, has recently been gaining attention.[18]

I traveled to St. Thomas in the U.S. Virgin Islands to learn about the paradoxically named "island vodka" dreamed up during a hurricane by Chef Todd Manley. A large cruise ship was in port the day I walked into Bar + Kitchen in the tourist district of Charlotte Amalie, the capital. Chef Todd stood in a doorway framed by two large, potted breadfruit trees, waiting to greet me. We took seats at a high-top table in the restaurant's outdoor courtyard to talk. "I've been accused of being overly passionate," Todd said by way of a warning, and then launched into his story.

When Hurricanes Irma and Maria hit the Virgin Islands within just two weeks of each other in 2017, Todd was running a restaurant on St. Croix. The electricity was out. Todd and his two business partners were in the hot and humid darkness, without customers, without revenue, without plans. Their conversation, born out of boredom and a need for business, shifted toward new product ideas. Todd mentioned that he had been experimenting with guinea grass, an invasive species in the Virgin Islands, as the basis for a new distilled alcohol. Todd liked the idea of helping to eradicate a troublesome weed but didn't like the taste of the beverage that resulted from his efforts. The three chefs cast about ideas for other crops that might be fermented and distilled. Someone mentioned breadfruit, but Todd kept steering the conversation in other directions. But the idea stuck with one of Todd's partners, who brought the talk to a close: "Todd, why don't you just get breadfruit and be done with it. You know it's going to work." "An hour later," Todd told me, "we were mashing breadfruit."

The first batch of Mutiny Island Vodka was homemade: breadfruit mash fermented in a five-gallon plastic bucket and distilled over a gas burner. Todd doled out shots to several of his friends, mostly chefs, to gauge their opinions. Most thought the flavor was too strong. The U.S. government's *Beverage Alcohol Manual* defines vodka as "neutral spirits distilled or treated after distillation . . . so as to be without distinctive character, aroma, taste or color." In other words, the taste of vodka should be like its color, clear, not reminiscent of the fruits or grains that made it. Todd packed some of his first batch along on a trip to Hawai'i and poured a sip for his friend, Sam Choy—the celebrity chef who would later author the 2022 *'Ulu Cookbook*—without telling him what the drink was made from. According to Todd, Sam closed his eyes for a long while after tasting then asked, "*'Ulu?*"[19]

While we were talking at his restaurant in St. Thomas, even though the original batch had been made more than six years earlier and the supply was dwindling, Todd disappeared into the restaurant's kitchen for a moment, then brought out a dusty bottle that still contained a few inches of nearly clear liquid. He filled a shot glass less than halfway. I took a whiff and then a sip. It smelled like Earl Grey tea and tasted, well, like breadfruit. Since 2017 Todd has been adjusting the recipe to make the taste more neutral. "I like the original one better," he told me, but his customers prefer the clear, clean taste of today's version. The packaging even touts the simple ingredients: "Breadfruit and Caribbean rainwater." The water used to produce that first batch was hurricane rain, evaporated from the Atlantic Ocean off the coast of Africa and showered in torrents over the Virgin Islands.

"When I hear 'vodka' I think of cold, gray cities in Russia," I confessed to Todd as we moved our seats under the shade of a large umbrella to escape the piercing tropical sunshine. "Mutiny is different," he explained, "It's not vodka, it's *island* vodka: a new drink category." And it can represent any island where breadfruit grows. Most Mutiny Island Vodka is made from breadfruit grown in Indonesia and imported to the Virgin Islands. But Todd showed me bottles stamped "Puerto Rican Reserve" that prominently featured La Monoestrellada, the lone-starred flag of the territory. "This is made from 100 percent Puerto Rican *pana*," he told me. Todd claims he can taste the subtle differences in vodka made from different islands' breadfruit. His plan is to capitalize on that terroir by creating unique single-sourced batches, each representing a specific place where breadfruit grows. Barbados, the Dominican Republic, Jamaica, Montserrat, Tahiti. Hawai'i, obviously.

Todd Manley views his mission as more than simply producing liquor. His concept for Mutiny Island Vodka combines the rebellion of the mutiny on the *Bounty* with the against-the-grain nature of radical environmentalism. Breadfruit, to Todd, seems like the worthiest ally in efforts against climate change and other kinds of environmental degradation. Mutiny Island Vodka's website touts its commitments to sustainability. "Better cocktails. Better planet" is one of Todd's trademarked phrases. Todd speaks of the *Bounty* mutineers as though they were forerunners to the champions of today's sustainability and social justice movements, resisting the environmentally degrading and dehumanizing effects of the sugar plantation system and its reliance upon enslaved labor through their refusal to participate in Captain Bligh's mission. In targeted advertising messages, regular drinkers of Mutiny Island Vodka are addressed as "mutineers."

"The one thing I want, when I'm dying," Todd said, "is to know I made a little dent in the way the world is going." And in what area, particularly, does Todd want his dent to be made? "Climate change. It's real; it's happening at an incredible pace." In the food and beverage industry, Todd sees his best opportunity to make a positive impact in his ability to promote certain crops, those grown sustainably, over those that aren't. He turned, addressing anyone within earshot, which was mainly the tourists passing by his restaurant, and declared, "Guys, you have got to plant breadfruit trees." If people will keep planting the trees, chefs like Todd will continue to come up with new culinary and libationary creations that make use of the fruits.

I was curious to see what a breadfruit-based menu would look like in a country where breadfruit was still novel, somewhere where its history did not go back centuries like the Caribbean or millennia like the Pacific. I've occasionally seen breadfruit listed as an ingredient in recipes that clearly arise from cuisines separate from these two world regions. Most tend toward Asian-fusion or Indian styles, which are connected to other early dispersals of breadfruit trees moving north and west out of its Pacific place of origin. I got the chance to meet a chef doing truly innovative work, inspired by breadfruit's most recent and still in-progress third global journey, while I was traveling in Ghana, where breadfruit was just starting to be taken seriously, following the import of hundreds of young, donated saplings in 2011.

Waiting in the lobby of a fancy Accra hotel for our appointment, Kwesi Agwani, the extension officer I was shadowing, and I felt out of place, having come straight from the muddy fields of his district. Everyone around us had on clean shoes and ours were caked with the red dirt of the farms we had inspected. Chef John, the man we'd come to see, arrived fashionably late, and his crisp pinstripe suit did nothing to improve Kwesi's and my feelings of inadequacy. The chef ordered us all mineral waters from the bar and began to extol breadfruit by the numbers. "Four thousand people have tasted breadfruit because of me. Ninety percent of what we do with potato, we can do with breadfruit." He reached into his jacket pocket and took out *two* mobile phones. "My phones are blowing up. Everyone wants breadfruit." Inclining both screens toward me, John scrolled slowly through his text messages. Indeed, many were variations of "Got any breadfruit?" accompanied by shrugging emojis of various skin tones.

John Jurai Oduro is the head chef at the Ghanaian Central Bank. The government employs him to prepare meals for the bank's staff and its international visitors daily. He explained to me that his goal is to offer his culinarily refined clients new twists on familiar haute cuisine. Breadfruit is among his favorite ingredients because of its versatility. "If it's unripe, I boil it like a yam," he explained. "If it's ripe, that's when I go into my beast mode and just see what I can make." Chef John, in beast mode, has turned ripe breadfruit into dishes ranging from croquettes to ice cream to baby food. He owns a nine-acre farm on which he recently planted seventy young breadfruit trees, but until those trees start producing, getting enough breadfruit is his main concern.

"I don't want to hang my reputation on breadfruit if the supply isn't there," he said, eyeing Kwesi, who looked away sheepishly. Before leaving Cape Coast, we were supposed to have collected forty breadfruits to bring to Chef John in the capital. But almost all the fruits we had gotten at the farms had found their way into extension workers' rucksacks or onto the mayor's desk; we had nothing for the chef. His disappointment was obvious. When our appointment ended, Kwesi and I made our way back to the cluttered comfort of his muddy SUV. Driving out of the hotel parking lot, Kwesi said to me, or maybe just to himself, "Ghana has got to plant more breadfruit."

CHAPTER XIII

Pushing Latitude

The National Statuary Hall inside the U.S. Capitol in Washington, DC, houses two statues from each of the fifty states. The statues, according to the 1876 act by which Congress established the colleciton, are to represent "deceased persons . . . such as each State may deem to be worthy of this national commemoration." Statuary Hall's roster includes President George Washington of Virginia, disability rights advocate Helen Keller of Alabama, inventor Thomas Edison of Ohio, and, ironically, another nation's monarch, King Kamehameha I of Hawai'i.[1]

Until recently one of the two statues representing Florida, the state of my birth, depicted Confederate General Edmund Kirby-Smith. Rightly, his was replaced in 2022 by a statue of educator and civil rights activist Mary McLeod Bethune. Not only is Bethune's the first statue of a Black American to be included in the Capitol's collection, but its installation represented, in the words of then-House Speaker Nancy Pelosi, "trading a traitor for a civil rights hero." The state's other statue is of someone, at least in Florida terms, equally heroic: John Gorrie, inventor of the mechanical refrigeration technologies that would later be used in air-conditioning and ice-making. Gorrie is the man who arguably rendered Florida habitable to the millions who live there today. And while author Marjory Stoneman Douglas has criticized users of air-conditioning as having "only a peripheral relationship with Florida itself," most Floridians gratefully raise a glass—a *chilled* glass—"to the man who made the ice, Dr. Gorrie."[2]

Florida is hot and humid. In Florida a good parking spot is defined by its shade, not just its proximity to the Publix. Children who grow up in Florida learn how to walk barefoot across a hot parking lot at the beach: you totter on the albedic painted lines like balance beams to save your feet from scorching on the black asphalt. In the afternoon heat of a Florida summer day, in the words of singer-songwriter David Dondero,

> The humidity—it's thick, you can cut it with a knife.
> If you'd like to take a breath here, honey,
> I'm gonna cut you out a slice.

On a day like that, you'd be hard-pressed to convince any farmer or gardener south of Apalachicola that it could ever get too cold to grow anything. Dondero sang of "purple skies and orange moons," how "plants are confident in June," but if you travel in January or February through the citrus groves and strawberry fields of central Florida, into the subtropical gardens on the edge of the Everglades, you'll find the confidence of both plants and planters waning on the rare but regular nights of winter frost.[3]

To help farmers and gardeners know what grows where, the Agricultural Research Service of the U.S. Department of Agriculture, in collaboration with Oregon State University, has developed a Plant Hardiness Zone Map that charts thirteen separate planting zones across the United States, based upon average low temperatures. Each zone is divided into two subzones, *a* and *b*, so there are really twenty-six zones altogether, each representing a temperature band of 5°F (a range of about 2.8°C). The coldest zone, 1a, is in Alaska and the warmest, 13b, is in Puerto Rico. Within the continental United States, the southern counties of Florida rate the warmest, with zones from 10a to 11b represented. That shift from 10 to 11 makes all the difference in the world to tropical species like breadfruit because the freezing point of water (0°C or 32°F) lies within zone 10a. This means that, in an average year, locations within that zone will see temperatures drop below freezing on some winter nights. In higher-numbered zones, 10b and 11, freezes are increasingly less likely.[4]

Horticulturist Julia Morton cautioned in her book *Fruits of Warm Climates* that breadfruit is not just tropical but "ultra-tropical," meaning that it has almost no tolerance for cold temperatures. Pat Breen, a retired Oregon State University horticulturist, has recommended that breadfruit be planted only in Zone 11 and above. The National Tropical Botanical

Garden has advised that breadfruit grows best between 21° and 32°C (about 70° to 90°F) and that temperatures below 5°C (41°F) can damage the leaves, causing them to curl and drop, depriving the trees of photosynthesized energy and eventually killing the tree. Jonathan Crane, a Miami-based agricultural extension officer told me that breadfruit trees can be damaged by temperatures as mild as 20°C (68°F) if the "cold" spell, by South Florida standards, lasts long enough. While it thrives in Hawai'i, the only places in the mainland United States where breadfruit should be able to grow are along a thin coastal strip of extreme southeast Florida around Miami in Zone 11a and throughout the Florida Keys in Zone 11b. An early twentieth-century nursery catalogue offered breadfruit saplings for sale, along with the warning that the species "will always be confined to the lower Keys of Florida." But that may be changing, especially if the efforts of a small, loosely affiliated, and globally dispersed band of radical gardeners—aided by a warming climate—are successful.[5]

It generally gets colder the further north or south you travel from the equator. This rule, however, is punctuated by exceptions. If latitude and temperature were exactly correlated, with no other factors at play, we wouldn't need the Plant Hardiness Zone Map; we could just paint a Mercator world map with horizontal bands of color representing the zones. But experience shows that places along a single line of latitude can experience vastly different climates.

I was born in Tampa, Florida, at 28° north latitude. Halfway around the world, also at 28° north latitude, is the summit of Mount Everest. The nearly incomparable climates of my beachside birthplace and the highest peak in the world, perpetually frozen at the edge of the stratosphere, differ not due to latitude but instead the myriad other geographical differences between these places, elevation most of all. As you move higher above sea level, you leave the insulating warmth of the earth's mass. You enter the thin air of the upper troposphere, where oxygen and heat energy are rare enough to make you gasp for breath as you shiver in the cold.

Higher elevations can mimic higher latitudes. This is why, since ancient times, wealthy lowland-dwellers have made summer retreats to second homes built in the cooler climates of nearby hills and why glaciers can still be found, at least for now, atop the equatorial mountains of New Guinea and Tanzania. In addition to elevation, other factors like wind and ocean currents, proximity to moderating water bodies and heat-producing cities, humidity, soil type, surrounding vegetation, albedo, and aspect—whether a location is north-facing or south-facing—contribute to the creation of microclimates: small isolates of one climate surrounded by larger areas of another. The concept of microclimates describes why tiny pockets of near-tropical climate can exist north of the line of Cancer or south of Capricorn, often at a scale too fine to be represented on the map of a country. But to the individual gardener, knowing and planting within the correct microclimate can mean the difference between a cold-sensitive plant's dieback and its survival.

Most American farmers and gardeners use the Zone Map to determine which plant species will thrive in their zones. Others view the map as a challenge. These extreme gardeners, or "zone-pushers" as they're sometimes called, use advanced techniques of horticultural hacking to exploit or create microclimates, imitating warmer zones around their tropical plants, hoping to coax them through each winter in direct defiance of the Zone Map. So when University of Florida horticulturist Herbert S. Wolfe wrote in his fifty-year retrospective on tropical fruit cultivation in Florida that "only the leader of a forlorn hope would still plant . . . the breadfruit" in the state, some heard a gauntlet hit the ground. Today's zone-pushers in Florida have three good reasons to be optimistic about their chances for success when it comes to growing breadfruit.[6]

First, the Zone Map is delineated at a geographical scale that many believe to be too coarse as it fails to take into account the subtleties of microclimates that an attentive gardener can find or create by paying close attention to the precise placement of their plants and other factors like wind protection and winter insulation with burlap, bubble wrap, and other blanketing materials. One zone-pusher I met strings Christmas lights around his cold-sensitive plants to provide a little incandescent warmth—along with some festivity—on cool nights.

Second, the genetic diversity of breadfruit is vast, and there may be certain mutations within the gene pool that could confer undocumented advantages of cold tolerance to individual trees planted outside the tropics. In 1977 the botanist Colin Leakey undertook a "breadfruit reconnaissance" tour throughout the Caribbean and remarked in his report that, "clearly, within the genetic range of the breadfruit, there is a much greater potential for adaptation to different climatic conditions than is indicated in the standard text book accounts." Leakey knew a thing or two about evolutionary adaptation, having been born into the famous family of paleoanthropologists and, in turning to botany, became in his own words, "the one the *National Geographic* doesn't know about."[7]

Third, and finally, the world is getting warmer. According to a 2021 U.S. government report, the seven hottest years on record were the immediately previous seven years: 2014–2020. Even the current version of the Zone Map, published in 2023, is expected to become rapidly outdated as the planet warms. Climatologists predict that the years of the coming decades will continue to break heat records. In many places, this record-breaking heat will be disastrous. Climate change is devastating for people and ecosystems around the world, directly causing sea level rise on the coasts, desertification inland, and the rapid melting of ice near the poles and in the still-glaciated peaks of the world's tallest mountains. But in a very limited way, the predicted effects of climate change could benefit certain people, in part by expanding the hardiness zones of certain tropical plant species like breadfruit into places like the southeastern United States.[8]

The desire to grow breadfruit in the United States is not new. In the late eighteenth century, before the *Bounty* even left England, Thomas Jefferson was aware of the mission to transplant breadfruit to the Caribbean and had an interest in bringing the tree to his country as well. In March 1789, Benjamin Vaughan of Jamaica wrote to Jefferson promising to send breadfruit seeds as soon as the ship he called "the bread-fruit vessel," arrived in Jamaica later that year. Vaughan was either unaware that the Tahitian cultivars of breadfruit the *Bounty* was after were seedless, or he was thinking of the breadnut that the French had been attempting to introduce to Saint-Domingue (now Haiti).[9]

A year later, Vaughn broke the news of the mutiny to Jefferson, who thanked him all the same, for his "readiness . . . to communicate to us the Bread tree." Jefferson wondered, though, "will any of our climates admit [its] cultivation?" and confessed, "I am too little acquainted with it to judge." Jefferson may not have fully understood breadfruit's intolerance for cold, but he was still willing to give it a try. He wasn't the only American statesman with such an interest. In 1795, just two years after breadfruit had arrived in the Caribbean, George Washington, then in his second term as president, received a letter from Fairlie Christie, a member of the Jamaican House of Assembly, who promised to send "3 Bread fruit plants in Baskets for your Excellency," knowing of "your Excellency's Wish to have a Bread fruit plant." Christie noted that, "if it will thrive in the Southern parts of America it will be a great Aquisition." Washington responded, thanking Christie and expressing his hope that, "with a little aid, [breadfruit] may be reconciled to the climate of my garden." Whether or not Washington ever received Christie's breadfruit is unknown. A few months later Washington would write from Philadelphia to the merchant and Virginia State Senator Thomas Newton Jr. to say that although he had forgotten the name of the ship bound from Jamaica, he hoped that the breadfruit and other plants might be sent to his plantation when they arrived. Newton replied to say that he had "made Inquiry after the plants" but found no sign of them, and the topic seems to have dropped from Washington's correspondence soon afterward.[10]

Jefferson, however, maintained an interest in the idea for years. In 1796, Alexandre Giroud, founder of the Société Libre des Sciences et des Arts (Free Society of Sciences and Arts) in Saint-Domingue, sent what he called "*quelques Graines du . . . Arbre à Pain*" (some seeds of the . . . Bread Tree), which may have come from breadnut, rather than breadfruit, or perhaps from one of the sporadically seed-producing breadfruit cultivars. Giroud, believing the seeds were breadfruit, sent the packet to Jefferson along with a long and optimistic letter stating that "this precious plant succeeds perfectly here," and expressing hope that it "will also succeed in the southern states of your Republic." Giroud wanted to be sure that, if breadfruit could be cultivated in the United States, *he* would be remembered as the one who introduced it: "May I one day learn that I am the one who gave this beautiful present to your homeland!" He went on to predict that Jefferson's own plantation, Monticello, might be "in 10 years, covered with the beautiful shade and the nourishing fruits of this precious tree."[11]

Jefferson acknowledged receipt of the seeds and thanked Giroud in May 1797 for "this valuable present," noting that he planned to "take immediate measures to improve the opportunity it gives us of introducing so precious a plant into our Southern states." Then, to match Giroud's sense of grandeur, Jefferson concluded with his own melodrama: "One service of this kind rendered to a nation is worth more to them than all the victories of the most splendid pages of their history, and becomes a source of exalted pleasure to those who have been instrumental to it."[12]

Jefferson wasn't exaggerating, though, when he promised to take "immediate measures." On the very same day that he thanked Giroud for sending the seeds, Jefferson wrote to a North Carolina planter named Allen Jones, stating that "having only seven seeds" he intended to send "two to each of the states of Georgia, S. Carolina, and N. Carolina, reserving one for Virginia." (Florida then was under Spanish rule; otherwise Jefferson probably would have sent seeds there too.) The North Carolina–bound seeds would be entrusted to Jones. While Jefferson had been "too little acquainted" with breadfruit back in 1789, to know whether it might grow in the United States, by now he was more confident, falsely so, owing to some misinformation he had received from Giroud. In his letter to Jones, Jefferson repeated Giroud's assertion "that Capt. Cook found that tree bearing fully in New Zealand in a colder temperature than that of London." To Jefferson, this left "little doubt it may be raised in our Southern states." But Jefferson and Giroud were mistaken. Cook's journals include observations of breadfruit in many places, but he did not claim to have seen it in New Zealand where, both historically and today, it does not grow. Historian Philippa Mein Smith has written poignantly that New Zealand "proved a graveyard instead of a garden for the standard Polynesian fare of coconut, breadfruit and bananas."[13]

Jones responded to Jefferson in August 1797, thanking him for "the tin box containing the seed of the bread tree mentioned in your letter," adding that he knew it would require special care to grow breadfruit in North Carolina. "If they could be raised for two or three years in a Greenhouse and then in the Spring of the Year turned into the full ground, I should make no doubt raising them." But he acknowledged his own disadvantage: "I have no Greenhouse, and must therefore do the best I can without one." Both men apparently enjoyed the challenge, however, as Allen's letter concluded with a statement with which Jefferson doubtlessly would have agreed: "Surely nothing can be so gratifying to an enlightened mind as adding to the felicity of your fellow Citizens by increasing the means of Subsistence." The seeds sent to South Carolina had also been well received. In August, Thomas Bee, a Charleston-based judge and president of the Agricultural Society of South Carolina to whom Jefferson had sent that state's allotted seeds, acknowledged Jefferson's letter, along with the "Seeds of the Bread Fruit Tree." Bee promised that "the greatest Attention will be paid to the raising [of] this Fruit if they once Vegetate."[14]

The fate of Jefferson's breadfruit seeds, and of any young trees that may have sprouted from them, is unknown. Given the climate of the Carolinas, Georgia, and Virginia during the late eighteenth century, we may assume the trees did not survive their first winter. The idea of growing breadfruit in the United States, though, was more persistent than the plants themselves. In 1822 Samuel Maverick, who would later go on to become a legendary southern politician, wrote to Jefferson—who by then had long since retired from government service—asking "would not the Tea plant & Bread fruit Tree be Valuable to those people who will indure the Long tedious warm Summers of Alabama?" Jefferson responded a few months later but, citing "Age, debility and decay of memory," declined to comment directly on Maverick's questions related to tea and breadfruit.[15]

Today, zone-pushing breadfruit growers in Florida are betting that Colin Leakey was right, the USDA cartographers are wrong, and Thomas Jefferson was ahead of his time. They want to prove that breadfruit is more adaptable, with broader genetic diversity than is commonly thought and, conversely, that the plant hardiness zones are defined too broadly, that fine-scale microclimates can be found, or made, where breadfruit will not only survive but thrive, produce fruit, and best of all be selectively propagated to produce a true subtropical cultivar. And while they have not yet reached Monticello, breadfruit trees are growing and fruiting further north than the Plant Hardiness Zone Map says that they should.

Based in Miami, the Rare Fruit Council was founded in 1955 by self-trained horticulturist Bill Whitman to promote the growing of tropical fruit in South Florida, taking, in Whitman's words "fullest advantage of this climatic opportunity." Whitman had become interested in tropical horticulture while working aboard cargo schooners in French Polynesia after the Second World War. He wrote that during this year he "had breadfruit as a staple diet" and after returning to Florida he "missed this native food." Whitman was not the first to introduce breadfruit to Florida—he would later write of trees already growing and fruiting on Key West and Key Biscayne when he first planted one—but he is credited for getting other Rare Fruit Council members interested in zone-pushing, generally, and breadfruit in particular.[16]

Whitman showed that it was possible to grow breadfruit and many other tropical plants at his experimental garden in Bal Harbour, a site fortuitously located on a narrow barrier island between the Atlantic Ocean and Biscayne Bay. These water bodies moderated the temperature, and a bamboo hedge helped protect against wind and salt spray. Whitman wrote that he "considers the breadfruit to be the most cold intolerant of all his introductions." From 1954 until his death in 2007, though, Whitman's breadfruit trees successfully and regularly bore fruit at his Bal Harbour garden.[17]

If you want to see Bill Whitman's botanical collection today, don't travel to Bal Harbour. The executors of Whitman's estate have not kept up the maintenance of his beloved tropical fruit trees and bushes; many have died or have been removed. Instead, you need to drive about fifty miles south and west to the agricultural district of Redland on the edge of Everglades National Park. In Redland, palm nurseries and fruit stands far outnumber the kinds of hotels and boutiques that abound in Miami Beach. Botanical gardens and farms are abundant. Driving into Redland at sunrise one April morning I watched broad-hatted farm workers disembarking from flatbed trucks, ready to begin a long day of fruit picking.

I had come to visit several breadfruit growers, starting with Rafael Santangelo, an attorney relocated from New York, with a passion for fruit. When Bill Whitman died, members of the Rare Fruit Council, including Rafael, respectfully took cuttings from his collection to preserve his life's work. In Rafael's greenhouse I found hundreds of potted plants, many with Whitman's name written in white ink on small, black tags alongside the plants' Latinized binomials. Once established, the trees would be moved outside and planted in rows on his ten-acre farm. "It's probably the world's largest collection of Whitman germplasm," meaning clonal offspring from plants Whitman cultivated, Rafael told me.

When I arrived, Rafael was working at a New York pace on his Florida farm. Speed-walking in high rubber boots from the greenhouse to the field and back again, hauling hose lines, cutting dead branches, piling brush, he was a moving target for my questions, which he answered in staccato responses as though he were being charged by the word. As we neared the side of the greenhouse he gestured toward another man whom I hadn't yet seen, standing bent at the waist, methodically checking the labels on potted saplings. "That's Jorge; he helps me out," Rafael said, before hurrying back inside the greenhouse. Jorge Zaldivar, a guava farmer and the son of Cuban immigrants, is also an amateur historian with a database-like memory. Jorge introduced himself in Spanish and I barely got out my *con mucho gusto* before he began his discourse. He seemed to have been waiting for the opportunity.

Jorge stood still and spoke as quickly as Rafael silently darted about his work. He mentioned names that I had read in back issues of the *Proceedings of the Florida State Horticultural Society*—Atkins, Clift, Fairchild, Kong, McNaughton, Morton, Whitman, Younghans—like they were his old friends. For those deceased or retired, he recommended readings. For those still active, he promised to set up meetings. Jorge paused his lecture when Rafael called out for us to help move pruned palm fronds from the base of a tree into a burn pile. We obliged, dragging the razor-edged limbs gingerly, even in gloved hands, while my history lesson recommenced.

Jorge Zaldivar comes across as a patient expert among flighty amateurs. Make that *mostly* patient. His forbearance is tested by tropical fruit growers who don't keep adequate records of their trees' provenance. "You don't ID trees," Jorge complained, "you ID cultivars. 'That's a jackfruit.' No: that's 'NS-1,' the *first jackfruit in the country*," meaning that the meter-high sapling in a pot right in front of me had been air-layered from a tree that was part of an unbroken line of clonal offspring from the first named cultivar of jackfruit imported into the United States. Like the pedigree of a purebred dog or the names of one's own great-grandparents, Jorge stressed, it's important to know each tree's ancestry. As he pointed out the breadfruit, jackfruit, and many other rarities in Rafael's greenhouse, Jorge took care to explain where each had come from. Not just directly, as in, "Rafael bought this tree from the Pine Island Nursery," but genealogically, tracing the sapling's ancestry back to the tree from which it had been "mossed-off," and, if he knew, the original specimen that had first been imported to Florida.[18]

We strolled over to the only breadfruit tree planted outdoors on Rafael's property. It was small—reaching a little higher than my waist—but heavily leaved. A drip irrigation hose ran right alongside it, and a royal palm stood uncomfortably close by. "What will you do when the breadfruit grows and gets into the space of that palm tree?" I asked. "Oh, the palm comes down," Rafael laughed. Royal palms are ubiquitous in South Florida; breadfruit trees need all the space they can get. I knew that any breadfruit tree planted in the ground, outdoors, on Florida's mainland is at risk of being damaged by the cold. This one, though, was planted in January and had already seen a few nights below 50°F (10°C). "What did you do to protect it?" I asked. "Nothing. That little tree is a warrior!" Rafael beamed.

As global temperatures rise, the need for breadfruit trees to be warriors against the cold will decrease. Bill Whitman presciently wrote in 1991 that "if our planet continues to heat up, as some scientists predict, possibly our descendants will face a climatically changed environment," and for ultra-tropical trees grown outside the tropics, "cold protection would no longer be required." For now, though, South Florida still sees cold temperatures in winter. In January 2010, the region experienced twelve days with average temperatures below 12°C (53°F) in Miami. This may seem mild to many readers, but the National Weather Service called the event "a historic cold snap of both duration and magnitude," even though the temperature in Miami never got below freezing—something that has only happened twelve times since 1895. While the occasional cold front or freeze doesn't change the long-term trajectory of the global climate getting warmer, each cold night can be fatal for Florida's breadfruit.[19]

Using projections based on climate models, a research team from Hawai'i in 2019 estimated the potential range where breadfruit may be able to grow fifty years into the future. Their model predicted that by 2070 the world will have 27 percent more land suitable for breadfruit cultivation than today. One region that shows the most potential for more breadfruit suitability is the southeastern United States. If that prediction comes true, today's zone-pushers of Florida may be remembered as the vanguard of breadfruit cultivation in the mainland United States.[20]

As I continued to spend time with Florida breadfruit growers, I was reminded of the closing words of Thomas Jefferson's 1797 letter, thanking Alexandre Giroud for the breadfruit seeds he sent from Saint-Domingue: "May . . . your name be pronounced with gratitude by those who shall at some future day be tasting the sweets of the blessing you are now procuring them." While today Giroud is rarely remembered within Florida's agricultural community, I had an opportunity to fulfill Jefferson's wish after a visit to the Kelsey L. Pharr Elementary School in Miami. My brother Andy accompanied me to the school to meet science teacher Sam Wims, who offered to guide us through the school's "food forest," which he had designed and established several years earlier.[21]

In a series of small plots snaking their way between the school's main building and its parking lots, trees bearing breadfruit and other tropical fruits provide shade, food, and hands-on experience for Mr. Wims's life sciences students. We were there in August, when the breadfruit trees were heavy with fruit. Sam gave us a tour, picking ripe fruit as we walked, and talking about the educational opportunities his students could have, year-round, through composting, planting, pruning, and harvesting the fruits grown right outside their classroom. As almost always happens when doing this kind of research, I left the meeting with armloads of fruit freshly picked from the trees.

Later Andy and I roasted several breadfruits atop a barbecue grill and prepared a ramekin of garlic butter for dipping. Even though these breadfruits were not descended from Giroud's "*quelques Graines du . . . Arbre à Pain*"—they were seedless, and anyway, there is no record of any of

Figure 13.1 Sam Wims harvesting breadfruit at the Kelsey L. Pharr Elementary School in Miami, Florida. (Photograph by Russell Fielding.)

Jefferson's seeds producing trees that survived—I felt it was important to acknowledge this early act of botanical diplomacy. As I peeled and sliced the first Florida-grown breadfruit I would taste, I paused to say *"merci, Monsieur Giroud,"* in Jeffersonian gratitude, for the good effort he made as an early forerunner to today's breadfruit growers of Florida.

Breadfruit is still a niche crop in the mainland United States. Most private growers maintain only a single tree, grown as a challenge, a curiosity, or, in the case of many members of Florida's Caribbean immigrant community, a reminder of home. I tried to imagine what a full-scale commercial breadfruit orchard in Florida might look like. To get a preview of what climate change may someday make possible, I drove south from Redland, past the last mainland outpost of Homestead, and began island-hopping over the bridges that connect the Florida Keys. My destination was Grimal Grove, a two-acre botanical garden on Big Pine Key and the largest commercial breadfruit operation in the continental United States.

Adolf Grimal, a mid-twentieth-century tropical fruit enthusiast and contemporary of Bill Whitman, established the grove that now bears his name in 1955. On his property, Grimal hewed out the limestone rock that underlies much of the land in the Florida Keys, filled the excavations with soil brought in from mainland Florida, and built extensive irrigation works. It was, in Bill Whitman's words, "a Garden of Eden," growing breadfruit alongside many other kinds of tropical fruit. When Grimal died in 1997, though, his grove quickly reverted to nature. Caretakers had been appointed, but their work was beset by the need for constant tending of the trees, as well as the occasional hurricane devastation. Weeds choked the delicate fruit trees, and people began squatting on the property. When the current owner, Patrick Garvey, first encountered the grove in 2011 it had amassed nearly a million dollars in fines for Monroe County code violations. Still, he saw potential. With a background in community organizing and food justice, Patrick imagined an edible landscape, tended collectively by volunteers who would share in the garden's bounty. He set about restoring the grove to realize that vision.[22]

Patrick opened Grimal Grove to the public in 2016. One of its first visitors—uninvited, to be sure—was Hurricane Irma the following year. After the storm, Grimal Grove was in need of restoration once again, having "lost 85 percent of our trees. . . . [due to] three feet of saltwater over the whole grove" as Patrick described to a local journalist. Of all the original fruit trees Grimal had planted, his breadfruits weathered the storm best.

Having witnessed its resilience and knowing that hurricanes were likely to become both stronger and more frequent with climate change, Patrick told me that he was "done planting anything but breadfruit." Grimal Grove, the best example of commercial breadfruit production in Florida, now features dozens of breadfruit trees, which produce fruit for local markets and air-layers for other growers to plant. Maybe someday the fields of Redland will be lined with breadfruit trees, planted in rows and laden with fruit, ready to be harvested for sale at local fruit stands and grocery stores, and for export to bigger markets. For now, though, growing breadfruit north of the Florida Keys is a horticultural hobby, an exercise in zone-pushing that sees setbacks whenever Florida is hit with an occasional winter cold spell.[23]

I began to wonder who had pushed the zone furthest. Where in the world could I find the most extreme breadfruit tree, the one growing farthest from the tropics? For an expert opinion I called Chris Rollins, a retired horticulturist who managed Homestead's Fruit & Spice Park until his retirement in 2014. Chris was intrigued by the question but warned me

Figure 13.2 Patrick Garvey at Grimal Grove in Big Pine Key, Florida. (Photograph by Russell Fielding.)

that I was chasing a moving target. Zone-pushing, he said, was really about getting a tree through as many winters as possible before it inevitably succumbs to the cold. Plants are confident in June. So are Florida zone-pushers. But when winter storms roll in from the Midwest, occasionally dipping as far south as Miami or the Keys, tropical plant growers are reminded of their latitude and their confidence withers like a cold-blasted breadfruit leaf.

Acknowledging that my search for the most extreme breadfruit tree was really seeking the tree growing at the highest latitude *right now*, Chris and I talked about microclimates, each of us picturing a mental map of the Florida peninsula, its barrier islands in particular. We decided to implement some rules. First, the tree would have to be planted in the ground, not in a pot. This was meant to disqualify any tree that could be moved indoors when the weather got cold. Second, it could not be under any kind of permanent roof like a greenhouse, although the use of temporary coverings on cold nights was acceptable. I had seen breadfruit trees growing in greenhouses as far north as Chicago and even London. You can grow bananas in Iceland with the right combination of glass walls and geothermal heat. We were interested in trees growing outdoors, facing the elements. The age of the tree was important too. Our third rule was that it had to have survived at least one winter. Finally, and most importantly, the tree had to have produced fruit.

Florida was the only place in the continental United States where we thought breadfruit might grow under these rules. California, despite its agricultural abundance, was immediately ruled out. The California Current brings cold water down from the coast of British Columbia; the mild summers it creates are a blessing to the state's vineyards and other temperate crops, but the current gives California a climate that never produces the consistent warmth needed by tropical fruits like breadfruit. Chris spoke of breadfruit trees he'd seen, or heard about, north of Homestead: Miami, Fort Lauderdale, Boca Raton, maybe even West Palm Beach. I told him I'd heard of one in Bokeelia, a small island community off Florida's Gulf Coast.

Our conversation briefly left the United States to follow the Equator and consider other possibilities around the world, outside the tropics. Northern Mexico: too dry. South America beyond the Tropic of Capricorn: no real historical link. Africa: north of the Tropic of Cancer, you're in the Sahara; south of Capricorn, we knew only of *Treculia africana*, a distantly related species commonly called "African breadfruit." The Arabian

Peninsula: too dry. The Mediterranean and Iran: like California, too cool. India: north of the Tropic of Cancer, you're in the Himalayas. China: not really part of the culture or the cuisine. Australia: well, what about Australia?

Australia is almost the size of the continental United States, but whereas the United States is divided into forty-eight mainland states, Australia has only seven major states and territories, each far larger than the typical U.S. state. Queensland, in northeastern Australia, is bigger than Alaska and more than twice the size of Texas. The state's mainland coastline stretches from Cape York in the far, tropical north to Coolangatta on the southern border with New South Wales. This is equivalent, in both distance and latitude, to a line stretching from Tampa, Florida, to Caracas, Venezuela. The Queensland coast, then, can be imagined as a Southern Hemisphere equivalent of the entire Caribbean, plus the southern half of Florida.[24]

Could there be breadfruit growing in Queensland, at latitudes comparable to those of Florida? I decided to go see. My plan was to start in the northern city of Cairns, where I knew breadfruit could grow, and proceed south until it could not. My first stop was Fruit Forest Farm, a little more than 90 miles (150 km) south of Cairns, where I spent a day with Peter Salleras, a commercial grower with an eclectic assortment of tropical fruits. When I arrived, I found a small group of farm hands washing rambutans in a tank of swirling water. Asking for Peter, I was given a set of very fruit-centric directions: "he's past the star apples and the jackfruits, in the pomelos." Following these directions led me to a tall, white-bearded man in a traditional Australian slouch hat, talking on his mobile phone. "I gotta go; there's a bloke here from America wants to talk about breadfruit," I heard Peter say before ending the call and extending a massive hand in greeting.

Peter drove me around in his farm truck, an old Hyundai with a corrugated metal roof and no windshield, through row after row of neatly pruned, leafy breadfruit trees. It was May, near the end of the harvest season, but many trees still had fruit. We picked a few breadfruits to bring back to the shed for lunch later, along with a jackfruit and some durian. Peter drove on to give me a tour of the entire farm. Stopping by a netted section of lemon-drop mangosteen bushes, Peter got quiet and told me to listen for the call of a cassowary: "he's in there somewhere." I got out of the truck to peer through the bushes, looking for the large, flightless bird. Suddenly two cassowaries appeared. It was a chick with an adult, the father, it would turn out, for as Peter explained, male cassowaries raise their young. I backed away quickly, but the young brown-and-black striped chick was curious and followed my every step. The father stood close by, watching, its teal and blue featherless head topped with a giant hornlike casque. I kept my eyes on the big male's feet, watching for any sign of an approaching kick. Cassowaries are often called "the world's most dangerous bird," due mostly to their powerful and sharply taloned feet. Peter and I were able to distract, or appease, the cassowaries by tossing them several of the mangosteens we had picked. Eventually the birds wandered back into the bushes, and we drove on, stopping at an irrigation pump and adding water to the truck's steaming radiator.[25]

Back at the shed where I'd first seen the rambutan-washing station, we sat down for lunch with several of the farm's employees. Jason, one of Peter's farm hands, quickly sliced a breadfruit into thick, crescent-shaped pieces and dropped them into a fryer. Peter opened a jackfruit and started passing around a wedge for us to pull pieces from. Trina, another of the farm's employees, cracked open a durian—we could all smell it before we saw it—and I took a thick piece of the custardy fruit to chew on. Finally, Peter produced a thick link of homemade macadamia nut and venison sausage, along with a large bush knife; we ate the nutty, meaty disks as quickly as he could cut them.

Dining on the produce of the farm, we talked about its uniqueness within the larger geographical setting. Even here at 17° south latitude, well within the tropics, Peter said the surrounding area gets too cold for some of the most ultratropical fruit species like breadfruit. Fruit Forest Farm sits on a slight ridge above a shallow valley to one side and Queensland's coastal plain to the other. Cold air, being denser than warm air, flows downhill like water, settling in the valleys and chilling the lowlands slightly more than the farm. If I was already in the difficult-to-grow-breadfruit zone of Queensland, it seemed hopeless to look for it at a more extreme latitude than I had found in South Florida.

A visit to one of Peter's acquaintances, horticulturist Roger Goebel, confirmed this suspicion. A little further down the coast from Fruit Forest Farm, Roger grows breadfruit along with a few other tropical species. He had previously done some research on the history of breadfruit in Australia and the current limits to its growth. Over tea and biscuits with homemade jam, Roger and I discussed his findings.

Breadfruit had been introduced to Australia through the Brisbane Botanic Gardens during the 1850s, but, since Brisbane's winters were too cold for breadfruit, the gardeners looked toward Queensland's tropical north for a suitable climate. The Australian market for breadfruit, as a food product, is driven mainly by Pacific Islander immigrant communities in the larger cities of Brisbane, Melbourne, and Sydney. In the early 2000s, Roger had conducted some reconnaissance similar to mine in Florida, looking for Australia's southernmost fruiting breadfruit tree. He found one in the town of Yeppoon at 23° south latitude. All the trees growing further south were either planted in greenhouses, potted, or hadn't produced fruit. Even the trees growing easily at Grimal Grove in the Florida Keys are at a more extreme latitude: further north than Yeppoon is south.[26]

In my search for the breadfruit tree growing farthest from the tropics, then, I had to rule out Australia. My focus returned to Florida, where both the human and environmental conditions seemed just right. The Gulf Stream brings tropical warmth up from the Caribbean, and the culture is heavily influenced by Caribbean and Latin American immigrant communities that know how to grow tropical fruits and what to do with them in the kitchen. With Chris Rollins's encouragement and contacts from his and Jorge Zaldivar's networks, I searched progressively north from the Keys on both sides of the Florida peninsula.

In Key West, I saw a massive breadfruit tree growing in an Old Town neighborhood. A couple walking by with their dog told me that the house behind which it grew belonged to Phil Crumbley, owner of an organic café and market nearby. Phil happened to be wearing a Hawaiian shirt adorned with images of ʻulu fruit and leaves when I dropped by the café unannounced. He shared with me an essay he had written, but never published, titled "Betsy and the Breadfruit Tree." It describes how the early settlers of Key West, as "a seafaring people," had brought back seeds and saplings from their voyages so they could plant breadfruit and other tropical trees throughout the island. These trees were not always valued, though, by the waves of later arrivals. "Some of the newcomers to Key West," the essay continues, "cut down exotic fruit trees to make way for a pool or driveway, not knowing of their previous past or purpose." When his neighbor, Betsy of the essay's title, asked Phil for his thoughts on her plan to cut down her own large breadfruit tree, his response was simply, "but Betsy, this tree feeds us." A goodhearted neighborly spat went on for a few years, with Betsy occasionally broaching the subject of cutting down the tree and Phil trying—by bringing her breadfruit-based dishes he had prepared—to show her the value of letting it stand. Two days after Betsy died in 1998, Hurricane Georges hit Key West and knocked the breadfruit tree down and onto her house. The one growing now in its place sprung from a root shoot after the storm.[27]

Near Phil's market, at the "Little White House," the winter residence of President Harry Truman, a large, perfectly pruned breadfruit tree grows at the side of the yard, its branches just reaching over the fence toward the sidewalk. According to Tom Whitney, the site's operations manager, the original landscaping plan for the property was developed in 1915, when the building was part of the Key West Naval Station. Willard Wells, the Navy's official gardener, regularly asked sailors to bring back to Key West plants that grow in the tropics. An avocado tree and a mango tree growing at the Little White House are both legacies of this effort. While the breadfruit tree is of more recent provenance—it came from Grimal Grove in 2019—its presence is in keeping with the spirit of the original horticultural

Figure 13.3 A breadfruit tree growing at the Harry S. Truman Little White House in Key West, Florida. (Photograph by Russell Fielding.)

"mission" issued by the Navy's top tropical gardener more than a hundred years ago.[28]

Back on the Florida mainland, in Homestead, I drove into a neighborhood of cookie-cutter houses and small, manicured lawns. "This doesn't seem like the right place," I thought, having gotten the address from one of Jorge Zaldivar's contacts. But when I arrived at the correct house number, Marie Ferraro stood in the driveway to greet me, wearing an apron from the Fairchild Tropical Botanic Garden. We walked around the house into the backyard and were instantly transported from a Florida suburb to a tropical Caribbean fruit forest. Two large trees, a breadfruit and a mango, stood in the center of the space, sheltering beneath their shared canopy a web of winding paths through plots planted with smaller trees and bushes. Ducks waddled along trails lined with orchids and plumerias, and a macaw screeched in its cage nearby. Marie explained that she had come to Florida from Jamaica, she was a great-grandmother, and this was her second breadfruit tree. The first had been blown down during Hurricane Irma, in 2017, and she planted this one from an air-layer taken from the original before it died. Root shoots regularly spring up, which Marie cuts, pots, and sells to other growers. The fruits that the tree produces, however, are not for sale. "I try to eat everything I grow," she said.

The northernmost tree I found that met our criteria—outdoors, planted in the ground, and bearing fruit—was in Loxahatchee Groves, just inland from West Palm Beach, at 26° north latitude. This tree's grower, Bobby Biswas, told me that it had survived three winters. When a local journalist wrote about Bobby's tree, her story made the front page of the *Palm Beach Post*. This one seemed like a contender for the "most extreme breadfruit tree" title.[29]

But then I remembered Bermuda. Out in the Atlantic Ocean, Bermuda sits at a latitude between that of Savannah, Georgia, and Charleston, South Carolina. Far beyond the tropics, well above Florida, and—based on conventional wisdom—much too far north for breadfruit to grow. And yet, there's tropical Bermuda. Like a Caribbean island that dragged its anchor and got carried northward by the Gulf Stream, Bermuda favors both the climate and the culture of its West Indian counterparts. Most tourist literature starts off by explaining that Bermuda is not, in fact, located in the Caribbean, because you wouldn't know if you weren't told. I wondered whether, on this pseudo-Caribbean outpost, I might be able to find an out-of-place breadfruit tree.

Figure 13.4 Marie Ferraro in her suburban garden in Homestead, Florida. (Photograph by Russell Fielding.)

Even before my plane touched down, the colors I saw through the window told me that things were different in Bermuda than any other place with which it shares a latitude. The sea was the same electric blue that you'd expect in Barbados or the Virgin Islands, but the foliage was reminiscent of early fall in North America—the leaves of deciduous trees appeared to be changing to their autumn colors. The buildings I could see from the air were whitewashed or painted in a range of pastels that recalled the Caribbean islands of Sint Maarten or Curaçao. Male officials working at the airport wore light linen blazers and the famous Bermuda shorts, which reached just to the tops of their knees, above their high socks and dress shoes. A place so warm that shorts were part of men's formal attire? This seemed promising for my extreme breadfruit quest.

Unlike the Caribbean islands, Bermuda's breadfruit history does not reach very far back and doesn't intersect with the history of slavery. Captain Bligh, Joseph Banks, and Voltaire all seem to have dismissed this Atlantic archipelago. It wasn't until 2015 that an amateur horticulturist named Paul Hollis managed to convince the notoriously strict Department of Environment and Natural Resources to allow him to bring one hundred tiny breadfruit saplings into Bermuda for distribution. They went quickly, being planted in pots and in backyards, on farms and on windowsills. Some of the trees have since grown large enough to have allowed air-layering and the creation of a second generation of Bermudian breadfruit. One of these would surely turn out to be the zone-pushing champion.

Shortly after arriving I met up with Abayomi Carmichael, a local engineer by trade, but whose passion was the cultivation of tropical fruit trees. For the next several days, Abayomi led me on an island tour, visiting other botanical enthusiasts, many of whom grew breadfruit as part of their tropical repertoires.

One grower, Carlos Amaral, took us deep into his mixed-fruit orchard to find the single breadfruit tree he had planted there. We meandered slowly, snacking on tree-ripened grapefruit and peaches along the way. When we reached the breadfruit tree, though, snack time ended. Carlos told me that his tree had not yet fruited, but he had hope.

I would soon learn that all the breadfruit trees in Bermuda were yet to bear fruit—except one. A Jamaican immigrant named Denton Jarrett showed Abayomi and me the largest breadfruit tree I would see on the island. Tucked carefully behind a homemade windscreen, the tree had produced modest yields of fruit for the past few years, a bounty Denton was

Figure 13.5 From left to right: Denton Jarrett, Abayomi Carmichael, and Frances Eddy at Jarrett's farm in Warwick Parish, Bermuda. (Photograph by Russell Fielding.)

happy to receive, since roast breadfruit was a delicacy from his childhood that he'd done without for quite some time. I snapped a photo of Denton standing in front of his fruiting tree and checked the metadata on my phone to see the latitude: 32° north. To the best of my knowledge, it's the most extreme breadfruit tree in the world. At least for now.

CHAPTER XIV

Two Trees

> And the Lord God planted a garden eastward in Eden;
> and there he put the man whom he had formed.
> And out of the ground made the Lord God to grow
> every tree that is pleasant to the sight, and good for food;
> the tree of life also in the midst of the garden,
> and the tree of knowledge of good and evil.
>
> —GENESIS 2:8–9 (KJV)

'*Ulu* thrives throughout the range it claimed during its first global journey, which took it from the site of its domestication in New Guinea to the far reaches of the Pacific. Where *'ulu*'s role has diminished due to globalization or neglect, it is being restored, thanks to the work of the Breadfruit Institute, the Hawai'i 'Ulu Cooperative, the Scientific Research Organisation of Sāmoa, and others. Some Pacific Islanders, like the Marquesans and Pohnpeians, have never forgotten breadfruit's role in their histories and have maintained its status today. Others, like the Hawaiians, Samoans, and Tahitians, are heirs to an *'ulu* renaissance.

Breadfruit is a staple crop in the Caribbean. It was introduced there through its second global journey, which carried it from the Pacific to St. Vincent and Jamaica, and then throughout the islands of the region. It's been creolized and is on its way toward becoming indigenous. While some of the old guard like authors Jamaica Kincaid and V. S. Naipaul have held a grudge against its past, others like Dwight Forde of Yelluh Meat and Marisol Villalobos of Amasar have seized onto its new potential; their efforts to share the bounty that the *Bounty* was meant to deliver proceed to the beat of Chi Ching Ching's "Roast or Fry!"

But which tree—*'ulu* or breadfruit—will lead the way for the third global journey of this "superfood?" Will the tiny trees being shipped by the hundreds to food-insecure regions of tropical Africa, Asia, and Central America be clipped root shoots of Pacific *'ulu* or air-layers of Caribbean breadfruit? In the genealogy of food crops, to which ocean basin will breadfruit trace its ancestry? Does it even make a difference, so long as the trees are distributed widely to provide food and ecosystem services to new peoples and landscapes? Should we even think of ʻulu and breadfruit as different trees?

In the book of Genesis, the Garden of Eden was planted with two important and diametrically opposite trees: the Tree of Life and the Tree of Knowledge of Good and Evil. Adam and Eve were banished from the garden for eating the fruit of the latter tree. The former, it would appear from references in the last book of the Christian Bible, the Revelation, has been transplanted to heaven, where, watered by a stream proceeding out of God's throne, the Tree of Life yields "fruit every month," and bears leaves "for the healing of the nations." What became of the Tree of Knowledge of Good and Evil is not addressed in scripture.[1]

Theologian Mark Makowiecki, however, has presented a compelling new way to think about these two primordial trees. According to Makowiecki, the text could allow the trees to be thought of as not entirely separate from one another. That is to say, like ʻulu and breadfruit being two names for the same species, the Tree of Life and the Tree of Knowledge of Good and Evil might be thought of as two names "referring to one and the same tree." This way of thinking would allow for the intertwining of both trees' purposes, the idea that giving life and knowing good and evil might merely be two facets of the same tree's existence. The tree which, after Adam and Eve ate its fruit, caused "the eyes of them both [to be] opened," may be the very same tree whose fruit could give eternal life. In this shared arboreal symbolism, then, life quite simply *is* the knowledge of good and evil. The longer one lives, the more one comes to know the best and the worst that the world has to offer. Can a single tree carry this much responsibility?[2]

I believe breadfruit can. But only if we allow the tree to be both ʻulu and breadfruit. In his theological exploration of the Garden of Eden's trees, Makowiecki, as any honest scholar will do, presented and considered arguments counter to his central thesis. Specifically, he cited portions of the text that seem to refute his assertion that the Tree of Life and the Tree of Knowledge of Good and Evil should be imagined as a single tree. Rather than choose one interpretation, however, Makowiecki settled upon "a vacillating singular-plural dynamic" in which he invited his reader to "conclude that it is not a matter of there being one *or* two principal trees in the

Garden of Eden, but of there being one *and* two trees." Finally, in laying out the various threads of his reasoning, Makowiecki cautioned that "to grasp the full meaning of a story, one must follow the narrative path from its beginning to its end." To me, this advice seems applicable to more endeavors than just academic theology and I have endeavored to follow it in my travels around the world following the various narrative paths of the tree called both *'ulu* and breadfruit.[3]

As *'ulu*, this tree journeyed across the Pacific in support of a heroic character in the story, a nation of adventurous and capable voyagers setting out to find new, uninhabited lands. As breadfruit, it made another journey, this time to the Caribbean in support of the story's villain, a colonial empire built on the backs of enslaved laborers. Breadfruit's role in its third global journey, the continuation of its story, will depend entirely upon how its historical roles are read into the future.

European explorers in the Pacific projected Edenic imagery onto the breadfruit tree, imagining it as the Tree of Life in the region's food systems. To them, it stood to reason that if these Pacific peoples still had access to the Tree of Life they must have been exempted from the curse with which God sent Adam and Eve east of Eden. While the imposition of foreign ideals onto breadfruit has attracted well-deserved criticism, we can still sympathize with the concept. From its origin stories—recall Kū emerging from the parched earth of his Hawaiian farm as a life-sustaining *'ulu* tree—to its present productivity and resurgence onto Pacific food scenes, breadfruit seems worthy of the title "tree of life." And it's not only Europeans who have applied the biblical metaphor. A 1967 article in the *Micronesian Reporter* that discussed a worrying blight among the region's breadfruit trees was titled "The Tree of Life Is Dying." In 2013 and 2014 the Breadfruit Institute distributed more than seven thousand *'ulu* saplings throughout Hawai'i, each with a copy of a brochure that included instructions for propagation and care. The title of the brochure was "Plant a Tree of Life— Grow 'Ulu."[4]

But, as its Caribbean history makes clear, breadfruit also bears a deep knowledge of good and evil. The saplings that Captain Bligh delivered to St. Vincent and Jamaica grew into trees that stood by silently as men, women, and children were stripped of their clothing, their names, and their humanity; families were separated—mothers, fathers, and children sold like livestock and sent to other plantations and other islands never to reunite; and people were worked literally to death with nothing to show for their labors except a meager meal of provisions and a legacy of having given their lives for the enrichment of men who claimed to own them. In the Caribbean, breadfruit abetted one of the worst systemic crimes against both the earth and humanity that history recalls: colonial plantation-based monocrop agriculture powered by racist chattel slavery. To that crime, breadfruit has long been charged as an accomplice.

Since most breadfruit cultivars are seedless, and therefore propagated clonally, it's likely that some individual trees growing today on the islands of the Caribbean share identical DNA with their parent trees, or those trees' clonal descendants, growing in the Pacific. Recent research has even matched specific Caribbean cultivars with their Pacific counterparts. Genetically speaking then, if genes define identity, a Pacific 'ulu and a Caribbean breadfruit might be, in Makowiecki's theological terms, literally *one tree*. Consider the range of human interactions that that single tree would have witnessed. In the Pacific, it sailed along with heroic Lapita and Polynesian navigators as a leaf-wrapped sapling, arriving at a new island home to be planted and to provide, Silversteinesque, for nearly all the needs of a people whom it loved. In the Caribbean, so many nautical miles away, genetically identical clonal scions of that Pacific 'ulu arrived as potted root shoots aboard HMS *Providence* and bore witness to unspeakable cruelty in service of unrestrained capitalism. It's a wonder, given these pasts, that breadfruit in the Caribbean today could ever approach the culinary and cultural position of 'ulu in the Pacific. And yet, the young artisan chefs making fantastic 'ulu creations in Hawaiian cafés and Tahitian bistros find worthy counterparts among the *ital* vendors of Jamaica and the food-truck chefs of Puerto Rico.

In the Pacific islands you can eat 'ulu uncritically. There, it has nothing to be ashamed of, except perhaps the culinary ground lost to rice and other introduced foods and the agricultural acreage lost to coffee and other cash crops. In the Caribbean, breadfruit is more complicated. Jamaica Kincaid described how "Antiguan children sense intuitively the part this food has played in the history of injustice and so they will not eat it." Bogotá-based journalist Karim Ganem Maloof wrote that "breadfruit is best if you remove the heart." I was so intrigued by this statement that I wrote Maloof to ask if he meant anything metaphorical by it, and he assured me that his advice was purely culinary: the "heart," or the portion of the stem that extends into the fruit itself, is both tougher and more porous than the rest of the fruit's flesh and when breadfruit is fried, as it often is in Colombian cuisine, the heart tends to absorb the most oil, becoming greasy and heavy. But the suggestion to cut out breadfruit's heart may have meaning, even if unintentional, beyond the kitchen. Perhaps on San Andrés island, one of Colombia's Caribbean possessions and Maloof's childhood home, it is the only way to enjoy breadfruit with a clean conscience. Colombia participated in the Atlantic slave trade as enthusiastically as any of its Caribbean neighbors. Removing the heart of a breadfruit grown there may serve to expunge its guilt for these crimes.[5]

It may be worthwhile at a point such as this, with breadfruit in the early stages of its third major journey, going global as a tool for sustainable development and emerging onto culinary scenes as a "superfood," to consider which sapling, with which historical trajectory, will be planted in Ghana, Honduras, Kenya, Pakistan, and beyond. And as climate change fuels the planet's tropicalization, which trees will the zone-pushers of Bermuda, North Queensland, and South Florida establish at consecutively broken records for highest latitude? Which fruits will produce the gluten-free flour on its way to being sold in grocery stores throughout the Americas and Europe, and which will be used in the recipes at restaurants worldwide? As breadfruit is now embarking on its third global journey, which history will it carry along: a history of heroic provision or a history of cruel enslavement? And can those histories even be separated, or, since breadfruit and 'ulu are both one and two trees, with one must the other necessarily go?

For a better perspective on these questions, I needed the insight of someone who had developed a deep knowledge of both breadfruit and 'ulu, someone who knew this tree in both the Caribbean and Pacific contexts. Someone like Mark Milligan.

Born and raised on the island of St. Croix in the Caribbean, Mark, an artist, has lived on O'ahu since 2008. I met him through contacts at the Pōpolo Project, a Honolulu-based nonprofit organization whose work "redefines what it means to be Black in Hawai'i—and in turn what it means to be Black in the world." *Pōpolo* is a plant. Its English name is glossy nightshade, and it is one of the canoe crops that were brought to Hawai'i along with 'ulu. *Pōpolo* produces clusters of small, purplish-black berries that, along with the leaves, were—and still are—used in a variety of traditional Hawaiian medicines for ailments ranging from sore muscles to tonsilitis. Owing to the berries' color, the word *pōpolo* has entered Hawaiian slang as a sometimes-derogatory reference to people with dark skin. Like so many ethnicity-based slang terms, though, *pōpolo* is in the process of being reclaimed, and the Pōpolo Project is the most visible form of that reclamation.[6]

I met Mark for lunch at a café attached to a small organic grocery store near Pearl Harbor. Over spring rolls and *ʻuala*, purple Hawaiian sweet potatoes, he spoke of his own transplantation from the Caribbean to the Pacific and reflected on the challenge of adapting both his life and his art to this new island environment. Central to his success on both fronts, it seems, were the plants. "I hear echoes of the Virgin Islands here," Mark said, as he pointed out trees on a distant ridge line and native Hawaiian flowers growing in the café's landscaping. "An island will embrace you when you arrive," he told me, "the land first, then the flora." This is why Mark's early artworks from Hawaiʻi feature Hawaiian landscapes and plants so prominently. Island *people* embrace you later, he explained; that takes a while.

We talked about island landscapes and island cultures, and of course island food. In St. Croix, breadfruit held no place of honor in the cuisine of Mark's childhood. "Poor man's food," he called it. But, like many traditions, both culinary and otherwise, this once-disparaged provision is seeing a renaissance in the Virgin Islands. "That used to be negative, now it's endearing," he explained. Like the word *pōpolo* in Hawaiʻi, breadfruit in the Caribbean is undergoing a process of reclamation.

"All these things have such memory for me," Mark said, still talking about Hawaiian plants. "Memory," he repeated, "but not a navel connection." I thought back to the Pacific traditions of burying placenta and umbilical cords beneath saplings to mark a child's birth and to root that child permanently to the land. "I don't have the navel connection, the umbilical connection, to these islands like I do to St. Croix," Mark said, and paused, slowly packing away one of his spring rolls to take home. He looked out, past the small lanai where we were eating and beyond the shrubbery that had reminded him of his first island. His eyes seemed to settle on the mountains, the lower slopes of the Koʻolau Range, which itself is the remnant western edge of a once-massive volcanic caldera that has ruptured, eroded, and slipped into the sea. "I don't have the navel connection here—but my sons do. This island gave me my sons." Mark remembers the breadfruit of the Caribbean. His two sons, rooted from birth in Hawaiʻi, know the *ʻulu* of the Pacific.

Maybe the most important lesson I've learned through all the travel, conversation, reading, and meditation that comprised this research is that breadfruit and *'ulu* both are and are not the same tree. I came to this realization after observing the parallel roles of *'ulu* and breadfruit in Pacific and Caribbean societies, but it's corroborated by the genetic research that's found both DNA matches and evolutionary divergences between generations of trees growing in St. Vincent for more than two hundred years and those that have grown in Tahiti for over a millennium.[7]

Breadfruit, so named by a pirate, was conflated with bread and then despised for doing what fruit does: growing freely on trees. Then, after the false start of the *Bounty*, *'ulu* root shoots from the Pacific were transported halfway around the world for breadfruit's Caribbean introduction. When HMS *Providence* arrived at Kingstown harbor in 1793, breadfruit took on its darkest historical role as a tool in service of slavery. Considering these records, Jamaica Kincaid was absolutely right to say that in the Caribbean "breadfruit is not a food, it is a weapon." Breadfruit was used as a weapon to defend the institution of slavery when supply chain disruptions and social progress threatened to force emancipation. Then, when breadfruit had become ingrained in the cultures and cuisines of the Caribbean, the threat to cut down breadfruit trees was another weapon, another way to enforce subjugation. Whether breadfruit itself or the "breadfruit mentality" that its abundant productivity was thought to have inspired, this tree and its fruit were weaponized in the Caribbean against the indentured and the enslaved, the Indigenous and the African.[8]

But weapons can be repurposed. A sculpture installed outside the headquarters of the United Nations in New York depicts a man holding a hammer in one hand and a sword in the other. The hammer is held aloft, ready to strike, and the sword is being pressed hard against the ground, its blade bending under the strain. The work, cast in bronze by artist Evgeniy Vuchetich, is titled *Let Us Beat Swords Into Ploughshares*; it was presented as a gift to the United Nations from the Soviet Union in 1959. The title is a reference to a passage in the book of Isaiah, in which the prophet foretells a time when "they shall beat their swords into ploughshares, and their spears into pruninghooks: nation shall not lift up sword against nation, neither shall they learn war any more." Even a weapon as deadly as a sword can be repurposed into a plow: a tool of agriculture, a tool of nourishment. To beat a sword into a ploughshare is an act against violence, an act of antiviolence, an act of reclamation.[9]

Before breadfruit's reclamation in the Caribbean can be complete, it must reconcile its role in slavery. In this, history may prove to be its

Figure 14.1 Let Us Beat Swords Into Ploughshares, 1957, Evgeniy Vuchetich, bronze sculpture, 111 × 76 × 35 in. United Nations, New York, exterior ground, acquired 1959. (Photograph by UN Photo/JC McIlwaine.)

greatest asset, for breadfruit revealed its disloyalty to the cause of oppression early on. Its initial participation in the mutiny on the *Bounty*, being thrown from the decks into the sea rather than carried along on the voyage, made clear that this Tree of Life from Tahiti—an island that symbolized liberty in European thought at the time—would not easily align itself with the injustice of bondage. But injustice, as it always is, was wickedly persistent. And liberty, as always, must remain vigilant.

When the *Providence* arrived in St. Vincent, four years after the *Bounty* was supposed to, and proceeded to distribute its potted cargo there, in Jamaica, and beyond, agents of empire—both people and plants—set out to establish a new crop intended not for the enslavers but for the enslaved. "A wholesome and pleasant food to our negroes," breadfruit was called. Those whom the breadfruit was intended to feed knew, or cared, little about the mutiny that had delayed its arrival and instead saw a bland and odd-looking fruit brought from the other side of the world to maintain their condition of enslavement even as world events were challenging the continuation of such a system. Was breadfruit viewed as a fellow sufferer in "a system of multispecies forced labor," as literary scholar Hannah Rachel

Cole has suggested? Did the enslaved laborers in the Caribbean, upon first confronting this strange new food, consider that the tree that bore it might have suffered an oppression like their own, "people and plants alike . . . located in the wake of colonial transplantation?"[10]

Like so much else about the thoughts and feelings of the enslaved, history bears scant record other than to note that it took at least fifty years in the Caribbean for breadfruit even to be considered worthy of human consumption. What is known is that the descendants of the enslaved people for whom breadfruit was intended have taken the first steps toward reconciliation. In 2020, the Caribbean Community—a regional intergovernmental organization like the European Union, whose name is often abbreviated CARICOM—created a Ten Point Plan for Reparatory Justice, meant to initiate the process of international reconciliation demanded by the as-yet insufficiently addressed history of slavery.[11]

First among the actions for which it calls, CARICOM's Ten Point Plan requires "as a precondition the offer of a sincere formal apology by the governments of Europe." But, opponents of national apologies might counter, no one alive today in Europe has ever legally enslaved anyone, and no one living now in the Caribbean has ever been legally enslaved. The philosopher Charles Griswold has written extensively about forgiveness and the preemptory role that an apology must play in the process of reconciliation. Griswold also acknowledged that "to apologize is to accept responsibility" and has addressed the separateness of personal and institutional apologies. Griswold stated that with apologies such as the one CARICOM demands, "it is not individuals, but the abstract entity, whose guilt is imputed," explaining that "because the individuals in question are long-since deceased, in practice it is the abstract entity that is responsible." Even though every English, French, Dutch, and Spanish enslaver has long since died, England, France, the Netherlands, and Spain all still exist as "abstract entities" and may rightly apologize for their own national misdeeds. Even though every person who was enslaved on a Caribbean island has also died, their descendants might still entertain the idea of forgiveness.[12]

CARICOM's plan calls not only for formal apologies, but for tangible reparations to be made. Specifically, the plan demands that European governments contribute more fully toward the cultural, economic, educational, and public-health development of the Caribbean region. And although the Ten Point Plan does not promise forgiveness in return for the actions it demands, it does state its commitment to "national [and] international reconciliation." Human rights scholar Roy L. Brooks has written that atonement, which he defines as "apology plus reparations," and forgiveness are "the key ingredients of racial reconciliation." And while we must remain cautious of the ubiquitous calls for what philosopher Myisha Cherry has called "racialized forgiveness," I have hope that European apologies and offers of reparations, followed by Caribbean forgiveness, might produce sincere reconciliation between societies on both sides of the Atlantic.[13]

But what of the trees? To illustrate the vast differences in cognition, communication, and experience across species boundaries, the philosopher Ludwig Wittgenstein once famously proposed that "if a lion could talk, we could not understand him." What if a breadfruit tree could talk? If it could apologize? Could we understand it? Would we know it was trying to atone for its past? Could we forgive it? Or since, according to Brooks, atonement is "apology plus reparations," might it be more beneficial to look first for evidence of breadfruit's reparations, before trying to listen for its apology? If, like the authors of CARICOM's Ten Point Plan, we define reparations primarily as contributions toward the development of a region and its people, then we might benefit from a reflection upon the contributions of the breadfruit tree in the Caribbean.[14]

As a humble provision, breadfruit has provided nourishment in times of hunger, whether resulting from droughts, hurricanes, or systemic injustices. Now it can also be seen as an ingredient on the refined menus of nouveau Caribbean cuisine, bringing renown and monetary reward to the region's chefs and entrepreneurs. Its cultural significance is witnessed in the botanical gardens, festivals, national dishes, and even the music of the islands. The trees provide food, shade, habitat, and carbon sequestration in a region feared to be on the front lines of the effects of biodiversity loss and climate change. And at the end of their lives, Caribbean people and breadfruit trees meet one final time in the humble dignity of caskets made from breadfruit lumber.

Breadfruit is central to both the Caribbean's dark history and its bright future. By embracing breadfruit as they have done, the people of the Caribbean seem to have signaled their readiness to forgive its role in their past. Perhaps breadfruit's bountiful and selfless provisioning can serve as an example to European governments that wonder what kind of, and how much, reparation might be required of them.

In the Pacific, *'ulu* carries none of the Caribbean's dark history. Across Oceania, *'ulu* is a staple crop, sometimes neglected, often overlooked in favor of imported foods, but wholesome, good, generous, and without colonial baggage. It's everywhere. So ubiquitous, in fact, as to have become part of the background and rendered nearly invisible on some Pacific islands. And when something fades to the background, the temptation is to forget it, to neglect to show gratitude for all its gifts. Given the choice between benign neglect and malicious oppression, of course, I'll take neglect. But neither is fitting for a tree with so much to offer. Especially now, as breadfruit embarks upon the third of its great journeys, emerging in the online fruit markets, organic grocery stores, and restaurants of American and European cities, and simultaneously among the food-insecure regions of Africa, Asia, and beyond.

On Maui I met Scott Fisher, co-owner of 'Ulu Brothers Farm, whose own family tree is rooted in both Native Hawaiian and European ancestries. While serving in the United States Marine Corps, Scott was stationed at Guantánamo Bay, Cuba, and he has traveled within the Caribbean, East Africa, and other colonized places. His perspective encompasses a wide swath of the earth's tropical geography. Scott holds a PhD in peace studies and serves as an environmental consultant in Madagascar on a project focused on food security and soil conservation.

As we stood behind a shed in his windswept *'ulu* grove, discussing the potential for breadfruit to contribute to food-security solutions in Africa, our conversation veered into the cultural and historical differences between breadfruit in the Caribbean and *'ulu* in the Pacific. Scott made the suggestion, remarkable for its simplicity, that it should be Hawaiians or other Pacific Islanders who take the lead in global efforts to introduce breadfruit to new places. To extend Scott's idea, when organizations like the Trees That Feed Foundation introduce breadfruit into new regions, they should view their work as continuing *'ulu*'s Pacific journey, not breadfruit's Caribbean legacy. He pointed out that it might even be beneficial for there to be a second breadfruit introduction to the Caribbean, now more than two centuries after Captain Bligh and long past emancipation, ceremonially led by Pacific Islanders and intentionally free from overtones of colonialism and slavery. To me, it felt as though Scott's suggestion was to replace, on a global scale, breadfruit with *'ulu*. To uproot the Tree of Knowledge of Good and Evil and, in its place, to plant the Tree of Life.

This idea was hard to accept. I agreed that Pacific Islanders could and should have a larger role in *'ulu*'s third global journey. In some situations breadfruit has been represented only as a Caribbean plant, its Pacific origin diminished or ignored. For example, in Jonathan Drori's brilliantly illustrated and geographically organized book, *Around the World in 80 Trees*, breadfruit is listed under the "Mexico, Central America, and the Caribbean" heading, along with avocado, sapodilla, and lignum vitae—all species native to the American tropics—not under "Oceania." Similarly, at the Royal Botanic Gardens, Kew, the single breadfruit tree on display in the garden's Palm House is found in the building's central atrium, with plants from the Americas, not in the north wing with the rest of the Asian and Australasian species. Both choices can be explained: Drori's text clearly and accurately presents breadfruit's origin in the western Pacific, its history throughout the Pacific islands, and its transplantation to the Caribbean. The specimen at Kew was donated by the government of Trinidad and Tobago, justifying its placement with the other plants of the Americas. Whether in a book or in a garden, a single plant can only be in one place at a time. Still, it can sometimes feel as though breadfruit's Caribbean history overshadows the history of *'ulu* in the Pacific. And given the darkness of breadfruit's story in the Caribbean, it makes sense to wish for change, a diminishment of breadfruit and an elevation of *'ulu*, a turning away from the Tree of Knowledge of Good and Evil and an embrace of the Tree of Life.[15]

But I don't think the Caribbean experience with breadfruit should be diminished. When oppression is the problem, exclusion can't be the solution. Instead I envision an international, multiethnic breadfruit/*'ulu* ambassadorship comprising those who know the tree best from both regions. Pacific Islanders can extoll the life-giving nature of *'ulu*, its reliability in times of need, its abundance and steadfastness. Hawaiians can tell the story of Kū's sacrifice; Pohnpeians can show the world how *mahi*, freshly harvested breadfruit, brings a party together (especially when paired with *sakau*); and Marquesans can teach the method of making *ma*—that fermented paste stored underground as insurance against famine. People from the Caribbean can bring their own history of reconciliation with breadfruit, how it overcame and is overcoming its associations with colonization, empire, genocide, land degradation, oppression, and slavery to become—owing to the abundant generosity, both of the tree and of the people—a beloved creole food, rightly placed alongside the ackee and saltfish of Jamaica's national dish or into the oil-down pot of Grenada. Both 'ulu and breadfruit have something important to offer the world. Each offering is distinct, uniquely Pacific and uniquely Caribbean. To exclude either perspective would be to diminish breadfruit's potential.

These two trees, 'ulu and breadfruit, like Eden's Tree of Life and Tree of Knowledge of Good and Evil, might somehow be thought of as both "one *and* two trees." To botanists, they are one. Scientists classify both 'ulu and breadfruit as *Artocarpus altilis* and have only begun to trace the effects of terroir, the variation from cultivation, and the slight genetic divergence that's occurred over the last two centuries. Despite these differences, the fruit tastes equally delicious whether roasted atop a Jamaican coal pot or among the hot lava rocks of a Hawaiian *imu*. But in their historical experiences, 'ulu and breadfruit are two distinct trees. 'Ulu's teachings are different from the things we can learn from breadfruit. During its third global journey, as the trees are planted in new soils and the fruits adorn new tables, it is my hope that the lessons from both 'ulu and breadfruit may also be shared with the world. 'Ulu teaches us how to give gifts abundantly, even sacrificially; breadfruit shows us how to make amends, even for the worst of our wrongdoings. 'Ulu evokes gratitude; breadfruit invites forgiveness.[16]

Acknowledgments

A Jamaican proverb states that "the more you chop breadfruit root, the more it spring." Reminiscent of these resilient shoots rising from a chopped root, I've found that the more you study breadfruit, the more breadfruit enthusiasts and experts spring up with generosity to help you on your way. Among them, I wish to express my deepest thanks to the following.

First, in Hawai'i, *mahalo* to Failautusi Avegalio, John Cadman, Sam Choy, Joannie Dobbs and Alan Titchenal, Travis Dodson, Elias Ednie, Scott Fisher, Duane Lammers, Linda Larish, Noa Kekuewa Lincoln, Patrick Kirch, Ken Love, Anissa Lucero, Mark Milligan, Mike Opgenorth, Jackie Prell, Dana Shapiro, Tammy Mahealani Smith, Pōhaku Stone, Werner and Beatrice Thie, and Jack Turner. And, although I've never yet had the chance to meet her, Diane Ragone has inspired me through her life's work, just as she has many other breadfruit scholars and growers around the world.

Elsewhere throughout the Pacific I was fortunate to have learned from *'ulu* experts including Kerinina Leaupepetele, Chande Lutu-Drabble, Seeseei Molimau-Samasoni, and Tisa and Candy Mann in American Samoa and Sāmoa; Massimo Bianco, Paul Gardner, Roger Goebels, Stephen Kenney, Adrian Mitchell, Jenny Newell, and Peter Salleras in Australia; James Chee and Beko on Borneo; Hervé Ah-Scha, Ana Bakran and Tangy, Teva Beguet, Teihotu Brando, Emily Donaldson, Antonin Fioretti, Antonio Heitaa, Felicienne and Gabriel Heitaa, Beni Huber and

Thérèse Rattinassamy, Pevatunoa Levy, Tihoni Maire, Bryan O'Connor, Hinano Teavai-Murphy, and Collette Teikitohe in French Polynesia; Bob Bevacqua and Ajalyn Omelau on Guam; Billie Lythberg in New Zealand; Eugene Darsy and Johnny Silbanus on Pohnpei; and Kylie Hasegawa, Winnie Lee, and Julius Reyes on Saipan.

In and around the Caribbean my breadfruit teachers included George Johnson and Pericles Maillis in the Bahamas; Chris Alleyne, Sonia Blackman-Francis, Nathan Crichlow, Dwight Forde and Kim Hamblin, Barney Gibbs, Charlotte Prud'Homme, and Richard White in Barbados; Carlos Amaral, Abayomi Carmichael, Paul Hollis, and Denton Jarrett in Bermuda; Juan Carlos Rodriguez, Natalia Vallejo-Rivera, Marisol Villalobos Rivera and Jesús Martes Cordero, and Liannette Bezares, Edgardo Matías, and Ramfis Fuentes in Puerto Rico; Sébastien Martinon and Charles Moreau on St. Barth; Ashley Cane, Michael Gloster, Kynella Nichols, Vincent Reid, and Jerrol Thompson in St. Vincent and the Grenadines; Laura Roberts-Nkrumah in Trinidad and Tobago; and Julius Jackson, Todd Manley, Nate Olive, and Art Wollenweber in the U.S. Virgin Islands.

It was a pleasure to return to the state of my birth and to investigate the northward spread of breadfruit in Florida. There I learned from Larry Atkins, Ken Banks, Bobby Biswas, Crafton Clift, Jonathan Crane, Phil Crumbley, Marie Ferraro, Patrick Garvey, Arturo Gonzalez, Aleena Hayat, Brett Jestrow, Benoit Jonckheere, Louise King, Michelle Leonard-Mularz, Craig Morell, Rafael Santangelo, Cynthia Schaeffer, Edelle Schlegel, Matt Snow, Ivan Torres Hidalgo Gato, Tom Whitney, Larry Wiggins, and Jorge Zaldivar.

Breadfruit experts live all over the world and I did my best to find, visit, correspond with, and learn from as many as possible. I am grateful for the insights I gained from Richard John Lynn, Susan Murch, and Kenna Whitnell in Canada; the late Karim Ganem Maloof in Colombia; Estefania Vincenti and Paul Zink in Costa Rica; Samuel Glory Abbey, Victoria Abankwa, Kwesi Agwani, and John Jurai Oduro in Ghana; James Hanrahan in Ireland; Joseph Matara in Kenya; Nick Dekoning in Uganda; Nicholas Cronk, Mark Nesbitt, Jon Nicholls, Hellen Pethers, Diana and Michael Preston, Greg Redwood, Janet Portman, Anne Rainsbury, John Thieme, Nicholas Thomas, Jacek Wajer, and David Wootton in the United Kingdom; Javion Blake, Birgit Cameron, Hannah-Rachel Cole, Kerry Dore, Kate Fredericks, Doris Garraway, Alex Lando, Christine Lutz, Maia

Nuku, John Rashford, Chris Rollins, Elaine Savory, Spencer Segalla, Leith Steel, and Amy Tijong in the United States; and Lindsay Gasik, wherever she may be currently chasing durian.

Especially essential were the insights I gained from Mary and Mike McLaughlin at the Trees that Feed Foundation in Illinois. Many of the people I met both in person around the world and virtually came from contacts that the McLaughlins shared with me, along with Mary's advice that "only good people plant trees." After meeting so many people around the world whose professional lives were dedicated to tree planting, Mary, I think you're right!

Mary and Mike also introduced me to Michael Morrissey, a British-Jamaican author and educator living a semiretired life on the island of Bali in Indonesia. Michael has written the only other book in English of which I'm aware on the culture and history of breadfruit at the global scale. His *Breadfruit Stories* was published in 2021 by the Trees that Feed Foundation, and I highly recommend it. On my way home from Borneo in 2023 I stopped at Bali for several days to spend time with Michael. His health was failing but his hospitality knew no limits and he graciously welcomed me into his guesthouse. We had several long, breadfruit-centered conversations—with coffee in his open-air *bale* or over *nasi campur* at his favorite restaurants around Ubud. Michael died just a few weeks after my visit and I count myself fortunate to have gotten to know him.

At Coastal Carolina University, my research was supported by my colleagues in the HTC Honors College—especially Dean Sara Hottinger—as well as Stephanie Cassavaugh and Elizabeth Reed in the Office of Sponsored Programs and Research Services; Darla Domke-Damonte in the Center for Global Engagement; and Molly French in the Institutional Research, Assessment, and Analysis office. My students also provided valuable help to this project, including the archival research conducted by Madison Smith and the revisions suggested by the students in my Sustainability Capstone class. Funding for my initial research in Florida and Hawai'i was provided through the Professional Enhancement Grant program and I thank Provost Gibbs Knotts and his predecessors, as well as Rob Young in the Office of Research, for making these funds available. I am sincerely grateful to the Alfred P. Sloan Foundation—especially Vice President and Program Director Doron Weber and his staff, including Shriya Bhindwale, Ali Chunovic, and Swan Griffith—for providing the majority of the funding that supported the research for this book.

It was a pleasure to work with Jennifer Crewe, associate provost and director of Columbia University Press, as editor, along with her staff, including Alfha Gonzalez and Sheniqua Larkin. The reviews Jennifer commissioned of earlier drafts of this book pushed me to consider new aspects of breadfruit, its history and present role in the world's botanical, cultural, and nutritional landscapes and I am grateful to the reviewers for their insights, as well as to Gregory McNamee for his expert copyediting. Parts of chapter 7 were previously published as an article in the journal *Plant Perspectives* in 2024, and parts of chapter 13 appeared as an article in *Annals of the American Association of Geographers* in 2022. I appreciate the respective publishers' allowing me to reproduce some of the text here. I am grateful to those who allowed me to use their photographs in this book—Diane Fielding, Hannah-Marie Garcia, and Becky Hadeed—and to the songwriters, poets, and publishers who allowed the use of their lyrics. Aly Ollivierre of Tombolo Maps & Design created the aesthetically pleasing and cartographically accurate maps at the opening of each of the book's three major parts.

Finally, to my family—my lovely wife Diane and our children, Conrad and Margaux—thank you for traveling with me as much as possible and, when you couldn't make a trip, for understanding my absence. Each time they accompanied me to a new island for this project's fieldwork, Conrad and Margaux would play "Who can find the first breadfruit tree?" They got really good! It's for their curiosity, eagerness to explore, and willingness to develop into quite the young travelers that this book is dedicated to them.

Notes

Preface

1. Hans Sloan, *A Voyage to the Islands of Madera, Barbados, Nieves, S. Christophers and Jamaica* (London: B. M., 1707), 1:lii.
2. John Ellis, *A Description of the Mangostan and the Bread-fruit: the first, esteemed one of the most delicious; the other, the most useful of all the Fruits in the East Indies* (London: Dilly, 1775).
3. Chris Colin, "Can Breadfruit Save the World?" *Saveur*, June 28, 2016, https://www.saveur.com/can-breadfruit-save-world/. See also Richard Schiffman, "Breadfruit Is Here to Save the World," *Wired*, September 11, 2024, https://www.wired.com/story/breadfruit-caribbean-pacific-climate-change-super-food/. A valid counterpoint was made in a 2024 article posted on the sustainable food website FoodPrint, which conceded that "breadfruit is delicious, versatile and exciting," but, in an appeal to readers to consider both climate change and food justice seriously, argued that "it shouldn't have to save the world." Alicia Kennedy, "Some Foods Are Styled as 'Climate Saviors.' Who Are They Saving?," *FoodPrint*, May 21, 2024, https://foodprint.org/blog/some-foods-are-styled-as-climate-saviors-who-are-they-saving/.
4. Julia Flynn Siler, "'Food of the Future' Has One Hitch: It's All but Inedible," *Wall Street Journal*, November 1, 2011, https://www.wsj.com/articles/SB10001424052970203752604576645242121126386; Anay Mridul, "Breadfruit Could Be the Next Superfood, with Great Potential as a Plant-Based Protein, Research Finds," *Vegan Review*, September 24, 2020, https://theveganreview

[237]

.com/breadfruit-could-be-the-next-superfood-with-great-potential-as-a-plant-based-protein-research-finds/; "Rawlston Makes Stewed Chicken and Breadfruit," *Bon Appétit*, October 26, 2020, https://www.bonappetit.com/video/watch/from-the-home-kitchen-rawlston-makes-stewed-chicken-and-breadfruit; Nina Compton, "Breadfruit Pie," *Food & Wine*, April 15, 2025, https://www.foodandwine.com/breadfruit-pie-11712915; Colleen Vincent, "Fried Breadfruit," James Beard Foundation, https://www.jamesbeard.org/recipes/fried-breadfruit; Jessica Harris, "Soused Breadfruit," Food Network, https://www.foodnetwork.com/recipes/soused-breadfruit-recipe-1915896; Daphne Ewing-Chow, "Breadfruit Has All the Makings of a Global Future Food Trend," *Forbes*, September 30, 2023, https://www.forbes.com/sites/daphneewingchow/2023/09/30/breadfruit-is-a-climate-smart-superfood-with-global-appeal/; Kim Severson, "9 Predictions for How We'll Eat in 2024," *New York Times*, December 26, 2023, https://www.nytimes.com/2023/12/26/dining/food-trends-2024.html; Andrea Gibbons, "Jamaica Kincaid: Breadfruit, Gardens, Empire," *Writing Cities*, May 25, 2020, https://www.writingcities.com/2020/05/25/jamaica-kincaid-breadfruit-gardens-empire/.

5. As an environmental geographer, my main tools of research are the "time-honored ones of geographers in the field: notebook, camera, and map," and I employ the equally time-honored methods: "to observe, question, and . . . to participate." Any academic investigation involving human subjects must be conducted ethically and approved by appropriate oversight bodies. The methods used during the research that led to this book were evaluated and approved by the Institutional Review Board at Coastal Carolina University (protocols #2021.69 and #2022.71). John E. Adams, "Last of the Caribbean Whalemen," *Natural History* 103 (1994): 64–72, quote at 66.

A Note About Language

1. The spelling and capitalization of the term *CHamoru* here is in keeping with the standard established by Guam's "CHamoru Heritage Commission Act of 2016" (Public Law 33–236). See also *Utugrafihan CHamoru, Guåhan* (Hagåtña, Guam: Kumisión i Fino' CHamoru, 2024), available online at https://kumisionchamoru.guam.gov/utugrafihan-chamoru-guahan-guam-chamoru-orthography/. On the etymologies and variations of Pacific-language terms for breadfruit, see "*Kulu," *Te Māra Reo / The Language Garden*, https://www.temarareo.org/PPN-Kulu.html, and the sources cited on this profoundly well-researched website, especially Malcolm Ross, Andrew Pawley, and Meredith

Osmond, eds., *The Lexicon of Proto Oceanic: The Culture and Environment of Ancestral Oceanic Society* (Canberra: Australian National University, 1998), 1:127. A quick reference to the various terms for breadfruit used throughout the world can be found at "Breadfruit Species," National Tropical Botanical Garden, https://ntbg.org/breadfruit/about-breadfruit/species/.

1. The Giving Tree

1. The region here called *the Pacific* may also be referred to, with equal geographical correctness but longer linguistic histories, by names such as *Moana, Moana nui,* and *Te-moana-nui-o-Kiwa,* among others. Another term in common usage, which Pacific anthropologist Epeli Hauʻofa referred to as "a designation that I prefer above all others," is *Oceania.* Less helpful are the terms *Australasia* and *the South Seas,* and the tripartite framework consisting of *Melanesia, Micronesia,* and *Polynesia.* I have chosen mainly to refer to the region, its land and sea, as *the Pacific* and *Oceania,* owing to the broad familiarity, among readers of English, with those terms. The name *Pacific Ocean* (originally *Mar Pacífico*) was bestowed by the navigator Ferdinand Magellan and *Oceania* (originally *Océanie*) by the cartographer Adrien-Hubert Brué. For more on the naming and perceptions of the Pacific, see O. H. K. Spate, "'South Sea' to 'Pacific Ocean': A Note on Nomenclature," *Journal of Pacific History* 12, no. 4 (1977): 205–11; Epeli Hauʻofa, "Our Sea of Islands," *The Contemporary Pacific* 6, no. 1 (1994): 148–61; Epeli Hauʻofa, "The Ocean in Us," *The Contemporary Pacific* 10, no. 2 (1998): 392–410, quote at 403; Bronwen Douglas, "Geography, Raciology, and the Naming of Oceania," *The Globe* 69 (2011): 1–28; and Nālani Wilson-Hokowhitu, "Moana Nui Rising: A Response to Blue-Washing the Colonization and Militarization of 'Our Ocean,'" *The Contemporary Pacific* 35, no. 1–2 (2023): 115–23. On traditions related to the origin of ʻulu, see Caren Loebel-Fried, *Hawaiian Legends of the Guardian Spirits* (Honolulu: University of Hawaiʻi Press, 2002) and Kawehi Avelino, *No Ke Kumu ʻUlu / The ʻUlu Tree,* trans. Lilinoe Andrews and Kiele Akana-Gooch (Hilo, HI: Hale Kuamoʻo and Kamehameha Schools, 2008).

2. Raden S. Roosman, "Coconut, Breadfruit, and Taro in Pacific Oral Literature," *Journal of the Polynesian Society* 79, no. 2 (1970): 219–32, quote at 225; Martha Warren Beckwith, *Hawaiian Mythology* (Honolulu: University of Hawaiʻi Press, 1982); Annie Walter, "Notes sur les cultivars d'arbre à pain dans le Nord de Vanuatu," *Journal de la Société des Océanistes* 88–89, no. 1–2 (1989): 3–18; William Ellis, *Polynesian Researches, during a Residence of Nearly Eight Years in the Society and Sandwich Islands* (London: Fisher, Son, & Jackson, 1831), 1:69; Teuira

Henry, cited in Diane Ragone, "Ethnobotany of Breadfruit in Polynesia," in *Islands, Plants, and Polynesians: An Introduction to Polynesian Ethnobotany*, ed. Paul Alan Cox and Sandra Anne Banack (Portland, OR: Dioscorides Press, 1991), 203–20, quote at 216.

3. Shel Silverstein, *The Giving Tree* (New York: Harper & Row, 1964).
4. Mikaele Foley, "International Breadfruit Conference," *Polynesian Cultural Center*, November 2, 2018, https://www.polynesia.com/blog/breadfruit; Diane Ragone, "Farm and Forestry Production and Marketing Profile for Breadfruit (*Artocarpus altilis*)," in *Specialty Crops for Pacific Island Agroforestry*, ed. Craig Elevitch (Hōlualoa, HI: Permanent Agriculture Resources, 2011), 1–19; Amon P. Maerere and Elias R. Mgembe, "Breadfruit Production in Tanzania: Current Status and Potential," *Acta Horticulturae* 757 (2007): 129–34. The figure of seven hundred fruits per year was given in Anthony Julian Huxley and Mark Griffiths, *The New Royal Horticultural Society Dictionary of Gardening* (New York: Macmillan, 1992), 248, and may have originated in J. W. Purseglove, *Tropical Crops: Dicotyledons* (London: Longmans, 1968), 382.
5. Joseph Banks, *Journal of the Right Hon. Sir Joseph Banks*, ed. Joseph D. Hooker (London: Macmillan, 1896), 135. The original quote is from Virgil's *Georgics*, book 2, verse 458.
6. Genesis 3:17b,19a (KJV).
7. Banks, *Journal*, 135.
8. Glenn Petersen, "Micronesia's Breadfruit Revolution and the Evolution of a Culture Area," *Archaeology in Oceania* 41, no. 2 (2006): 82–92.
9. Oliver Sacks, *The Island of the Colorblind* (New York: Vintage Books, 1998), 60.
10. Louis Lewin, *Phantastica: Narcotic and Stimulating Drugs, Their Use and Abuse* (New York: Dutton, 1964), 223.
11. On the mysteries of Nan Madol see Christopher Pala, "Nan Madol: The City Built on Coral Reefs," *Smithsonian*, November 3, 2009, https://www.smithsonianmag.com/history/nan-madol-the-city-built-on-coral-reefs-147288758/; Louella Losinio, "Mysteries of Nan Madol," *Guam Daily Post*, August 15, 2016, https://www.postguam.com/forum/featured_columnists/mysteries-of-nan-madol/article_b9f86910-5140-11e6-8e3a-931df21db49d.html.
12. Oliver Goldsmith, *The Deserted Village—a Poem* (London: Griffin, [1770] 1927), 4.
13. Lewin, *Phantastica*, 224.
14. F. Raymond Fosberg and Marie-Hélène Sachet. *Flora of Maupiti, Society Islands*, Atoll Research Bulletin no. 294 (Washington, DC: Smithsonian Institution, 1987), quotes at 2.

15. E. S. Craighill Handy, *The Native Culture in the Marquesas* (Honolulu: Bernice P. Bishop Museum, 1923), 64.
16. Roger Neich and Mick Pendergast, *Pacific Tapa* (Honolulu: University of Hawai'i Press, 2005); Michele Austin Dennehy, Jean Chapman Mason, and Adrienne L. Kaeppler, "Breadfruit Tapa: Not Always Second Best," in *Material Approaches to Polynesian Barkcloth: Cloth, Collections, Communities*, ed. Frances Lennard and Andy Mills (Leiden: Sidestone Press, 2020), 61–69, quote at 62; Jennifer Newell, *Trading Nature: Tahitians, Europeans, and Ecological Exchange* (Honolulu: University of Hawai'i Press, 2010), 18.
17. Billie Lythberg, email to author, September 20, 2022.
18. Donald Kerr, *Census of Alexander Shaw's Catalogue of the Different Specimens of Cloth Collected in the Three Voyages of Captain Cook to the Southern Hemisphere, 1787* (Dunedin, New Zealand: University of Otago, 2015).
19. Alexander Shaw, *A Catalogue of the Different Specimens of Cloth Collected in the Three Voyages of Captain Cook, to the Southern Hemisphere* (London, 1787); Alexander Leitch, *A Princeton Companion* (Princeton, NJ: Princeton University Press, 1978), 261–62. On Hosack, see Victoria Johnson, *American Eden: David Hosack, Botany, and Medicine in the Garden of the Early Republic* (New York: Liveright, 2018).
20. Shaw, *Catalogue*, 7.
21. Val Krohn-Ching, *Hawaii Dye Plants and Dye Recipes* (Honolulu: University of Hawai'i Press, 1992).
22. Brien A. Meilleur, Noa Kekuewa Lincoln, Joannie Dobbs, C. Alan Titchenal, Richard R. Jones, and Alvin S. Huang, *Hawaiian Breadfruit: Ethnobotany, Human Ecology, Agronomy, Nutrition, Modernity* (Honolulu: College of Tropical Agriculture and Human Resources, University of Hawai'i-Mānoa, 2024); Noa Kekuewa Lincoln, Diane Ragone, Nyree J. C. Zerega, Laura B. Roberts-Nkrumah, Mark Merlin, and A. Maxwell P. Jones, "Grow Us Our Daily Bread: A Review of Breadfruit Cultivation in Traditional and Contemporary Systems," *Horticultural Reviews* 46 (2019): 299–384.
23. On Duke Kahanamoku, the famous Hawaiian surfer, swimmer, and five-time Olympic medalist, see the 2021 documentary *Waterman*, dir. Isaac Halasima (Sidewinder Films).
24. The quote "I saw the angel in the marble and carved until I set him free" is often attributed to Michelangelo, but no original source is known. The idea behind the quote may have originated with Michelangelo's description of sculpture as proceeding "by the force of removing" in contrast to painting, which "is made by the means of placing," in a 1549 letter to the poet Benedetto Varchi. See Gaetano Milanesi, ed., *Le lettere di Michelangelo Buonarroti* (Florence: Le Monnier, 1875), 522–23.

2. Charismatic Megaflora

1. Daniel Stone, *The Food Explorer: The True Adventures of the Globe-Trotting Botanist Who Transformed What America Eats* (New York: Penguin Random House, 2018); David Fairchild, "The Jack Fruit (*Artocarpus integra*, Merrill): Its Planting in Coconut Grove, Florida," Occasional Paper No. 16, *Florida Plant Immigrants* (1946): 5–14, quotes at 5; Nyree Zerega, Tyr Wiesner-Hanks, Diane Ragone, Brian Irish, Brian Scheffler, Sheron Simpson, and Francis Zee, "Diversity in the Breadfruit Complex (*Artocarpus*, Moraceae): Genetic Characterization of Critical Germplasm," *Tree Genetics & Genomes* 11 (2015): 77–91.
2. P. Francisco Manuel Blanco, *Flora de Filipinas: Según el sistema sexual de Linneo* (Manila: Santo Thomas, 1837); M. Mohammed and L. D. Wickham, "Breadnut (*Artocarpus camansi* Blanco)," in *Postharvest Biology and Technology of Tropical and Subtropical Fruits*, ed. Elhadi M. Yahia (Philadelphia: Woodhead, 2011), 272–89; Tessa McSwain, ed. "Breadfruit Species," National Tropical Botanical Garden, https://ntbg.org/breadfruit/about/species/.
3. Auguste Trécul, "Memoire sur la famille des Artocarpees," *Annales des Sciences Naturelles* 3, no. 8 (1847): 38–157; McSwain, "Breadfruit Species."
4. Jonathan Drori, *Around the World in 80 Trees* (London: Laurence King, 2018), 194.
5. George R. R. Martin, *A Game of Thrones: A Song of Ice and Fire, Book One* (New York: Bantam, 1996), 415.
6. Frances M. Jarrett, "The Syncarp of *Artocarpus*—a Unique Biological Phenomenon," *The Gardens' Bulletin Singapore* 29 (1977): 35–39.
7. Names of breadfruit varieties given throughout this book are formatted according to the guidelines developed by the Royal Horticultural Society and found in The Royal Horticultural Society Botany Advisory Services, *Recommended Style for Printing Plant Names* (Wisley, UK: Royal Horticultural Society, 2004). Diane Ragone, *Breadfruit:* Artocarpus altilis *(Parkinson) Fosberg* (Gatersleben, Germany: Institute of Plant Genetics and Crop Plant Research, and Rome, Italy: International Plant Genetic Resources Institute, 1997); National Tropical Botanical Garden, "Ulu Fiti," *Breadfruit Institute*, August 15, 2018, https://ntbg.org/wp-content/uploads/2020/02/ulufiti_factsheet_aug2018.pdf; Diane Ragone, "Farm and Forestry Production and Marketing Profile for Breadfruit (*Artocarpus altilis*)," in *Specialty Crops for Pacific Island Agroforestry*, ed. Craig Elevitch (Hōlualoa, HI: Permanent Agriculture Resources, 2011), 1–19; Nyree J. C. Zerega, "The Breadfruit Trail," *Natural History* (December 2003/January 2004): 46–51.
8. Mary Kawena Pukui, *ʻŌlelo Noʻeau: Hawaiian Proverbs & Poetical Sayings* (Honolulu: Bishop Museum Press, 1983), 176, 226.

9. G. Bourdy, P. Cabalion, P. Amade, and D. Laurent, "Traditional Remedies Used in the Western Pacific for the Treatment of Ciguatera Poisoning," *Journal of Ethnopharmacology* 36 (1992): 163–74; Lois Lucas, *Plants of Old Hawaii* (Honolulu: Bess, 1982); Patrick Vinton Kirch, *A Shark Going Inland Is My Chief: The Island Civilization of Ancient Hawaiʻi* (Berkeley: University of California Press, 2012).
10. Ying Liu, Paula N. Brown, Diane Ragone, Deanna L. Gibson, and Susan J. Murch, "Breadfruit Flour Is a Healthy Option for Modern Foods and Food Security," *PLoS One* 15, no. 7 (2020): e0236300.
11. Ragone, *Breadfruit*; Ragone, "Farm and Forestry."
12. National Tropical Botanical Garden, *Breadfruit Institute*, https://ntbg.org/work/institute/; Ragone, "Farm and Forestry," 4.
13. Robert Cecil Murray Wright, *The Complete Book of Plant Propagation: A Practical Guide to the Various Methods of Propagating Trees, Shrubs, Herbaceous Plants, Fruits and Vegetables* (London: Ward Lock, 1981). I have found examples of the term "mossing off" only in the context of agriculture in the southeastern United States, particularly Florida. The phrase seems to be unique to the region and may have originated in the state. See, for example, Allen H. Andrews, *A Yank Pioneer in Florida* (Jacksonville, FL: Douglas, 1950), 384, 497; T. J. Sheehan and Jasper N. Joiner, *Propagation of Ornamental Plants by Layering*, Circular 141 of the Florida Cooperative Extension Service (Gainesville: University of Florida Institute of Food and Agricultural Sciences, 1955), https://ufdc.ufl.edu/UF00084279/00001, 1; R. W. Johanson, "What We Know About Air Layering," *Proceedings of the 4th Southern Forest Tree Improvement Conference, Athens, GA, January 8–9, 1957*, 126–31, quote at 126, available online at https://rngr.net/publications/tree-improvement-proceedings/southern/1957; Sam Mase, "Exotic Lychee Business Spreading in Florida," *Tampa Tribune*, August 6, 1950; Geert-Jan de Klerk, Wim van der Krieken, and Joke C. de Jong, "Review: The Formation of Adventitious Roots: New Concepts, New Possibilities," *In Vitro Cellular & Developmental Biology—Plant* 35 (1999): 189–99; Bianka Steffens and Amanda Rasmussen, "The Physiology of Adventitious Roots," *Plant Physiology* 170 (2016): 603–17.
14. H. Frederic Janson, *Pomona's Harvest: An Illustrated Chronicle of Antiquarian Fruit Literature* (Portland, OR: Timber Press, 1996), 29; Frits Warmolt Went, "Auxin, the Plant Growth-Hormone," *Botanical Review* 1, no. 5 (1935): 162–82; Ken Love, Robert E. Paull, Alyssa Cho, and Andrea Kawabata, *Tropical Fruit Tree Propagation Guide* (Honolulu: College of Tropical Agriculture and Human Resources, University of Hawaiʻi-Mānoa, 2017).
15. Wright, *The Complete Book of Plant Propagation*; Puran Bridgemohan, Musa El S. Mohammed, Arjune Ramoutar, Kimberly Singh, and Ronell Bridgemohan, "Air Layering (Marcotting) of Breadfruit (*Artocarpus Altilis*)," *International Journal*

of *Research and Scientific Innovation* 3, no. 9 (2016); Christopher Menzel, *The Lychee Crop in Asia and the Pacific* (Bangkok: Food and Agricultural Organization of the United Nations, Regional Office for Asia and the Pacific, 2002).

16. Dan Koeppel, *Banana: The Fate of the Fruit That Changed the World* (New York: Hudson Street Press, 2008).

17. The term *ultimate wave* was applied to the surf break at Teahupo'o most famously in the IMAX documentary *The Ultimate Wave Tahiti*, dir. Stephen Low (K2 Communications, 2010).

3. The Treeness of Life

1. Nyree J. C. Zerega, "The Breadfruit Trail," *Natural History* (December 2003/January 2004): 46–51; Patrick Vinton Kirch, *A Shark Going Inland Is My Chief: The Island Civilization of Ancient Hawai'i* (Berkeley: University of California Press, 2012); David Lewis, *From Maui to Cook: The Discovery and Settlement of the Pacific* (Sydney: Doubleday Australia, 1977).

2. Evelyn W. Williams, Elliot M. Gardner, Robert Harris III, Arunrat Chaveerach, Joan T. Pereira, and Nyree J. C. Zerega, "Out of Borneo: Biogeography, Phylogeny, and Divergence Date Estimates of *Artocarpus* (Moraceae)," *Annals of Botany* 119 (2017): 611–27, quote at 625.

3. Lindsay Gasik, "A Guide To 8 Kuching Markets and Fruit in Sarawak Malaysia," *Year of the Durian*, December 16, 2017, https://www.yearofthedurian.com/2017/12/kuching-market-guide.html.

4. Anmol Bhatia, Jennifer R. Lenchner, and Abdolreza Saadabadi, "Dopamine Receptors," *StatPearls*, https://www.ncbi.nlm.nih.gov/books/NBK538242/.

5. Fong-Ming Chang, Judith R. Kidd, Kenneth J. Livak, Andrew J. Pakstis, and Kenneth K. Kidd, "The World-wide Distribution of Allele Frequencies at the Human Dopamine D4 Receptor Locus," *Human Genetics* 98 (1996): 91–101.

6. Chang et al., "The World-wide Distribution"; David Dobbs, "Restless Genes," *National Geographic* 223, no. 1 (2013): 44–57. See also Luke J. Matthews and Paul M. Butler, "Novelty-Seeking DRD4 Polymorphisms are Associated with Human Migration Distance Out-of-Africa After Controlling for Neutral Population Gene Structure," *American Journal of Physical Anthropology* 145 (2011): 382–89.

7. On Pacific voyaging canoes, see David Lewis, *We the Navigators: The Ancient Art of Landfinding in the Pacific* (Honolulu: University of Hawai'i Press, 1972), ch. 3.

8. On the environmental clues used in Pacific navigation, see Lewis, *We the Navigators*.

9. Kirch, *A Shark Going Inland*; Geoffrey Irwin, *The Prehistoric Exploration and Colonisation of the Pacific* (Cambridge: Cambridge University Press, 1994), 31.
10. Brian Diettrich, " 'Summoning Breadfruit' and 'Opening Seas': Toward a Performative Ecology in Oceania," *Ethnomusicology* 62, no. 1 (2018): 1–27.
11. Kirch, *A Shark Going Inland*.
12. J. C. Beaglehole, ed. *The Journals of Captain James Cook on his Voyages of Discovery*, vol. 3, pt. 1: *The Voyage of the Resolution and Discovery 1776–1780*. (London: Routledge, 2017), 279.
13. Wade Davis, *The Wayfinders: Why Ancient Wisdom Matters in the Modern World* (Toronto: Anansi, 2009), 35.
14. E. S. Craighill Handy and Elizabeth Green Handy, *Native Planters in Old Hawai'i: Their Life, Lore, and Environment* (Honolulu: Bishop Museum Press, 1991), 9.
15. Kirch, *A Shark Going Inland*, quotes throughout.
16. Kirch, *A Shark Going Inland*, 67; Patrick Vinton Kirch, email to author, May 2, 2021. See also Mark D. McCoy, Michael W. Graves, and Gail Murakami, "Introduction of Breadfruit (*Artocarpus altilis*) to the Hawaiian Islands," *Economic Botany* 64, (2010): 374–81.
17. Handy and Handy, *Native Planters*; Tessa McSwain, "Breadfruit Species," *National Tropical Botanical Garden*, 2019, https://ntbg.org/breadfruit/about/species/.
18. The story of the fish-producing breadfruit tree is found in many sources and is a popular motif for a famous art form from Palau: carved wooden storyboards. Succinct, though slightly differing versions of the story can be found in Diane Ragone and Harley I. Manner, "*Artocarpus mariannensis* (dugdug)," in *Traditional Trees of Pacific Islands: Their Culture, Environment, and Use*, ed. Craig R. Elevitch (Hōlualoa, HI: Permanent Agriculture Resources, 2006), 136–37; DeVerne Reed Smith, "The Palauan Storyboards: From Traditional Architecture to Airport Art," *Expedition* 18, no. 1 (1975): 2–17; and Karen Nero, "The Breadfruit Tree Story: Mythological Transformations in Palauan Politics," *Pacific Studies* 14, no. 4 (1992): 235–60.
19. Nero, "The Breadfruit Tree Story," 239.
20. Mary Kawena Pukui and Samuel H. Elbert, *Hawaiian Dictionary: Hawaiian-English, English-Hawaiian* (Honolulu: University of Hawai'i Press, 1971); Handy and Handy, *Native Planters*; Noenoe K. Silva, *Aloha Betrayed: Native Hawaiian Resistance to American Colonialism* (Durham, NC: Duke University Press, 2004).
21. Chantal Spitz, *Island of Shattered Dreams*, trans. Jean Anderson (Wellington: Huia, 2007), 23–24.
22. Spitz, *Island of Shattered Dreams*, 24.
23. Spitz, *Island of Shattered Dreams*, 24.

24. Solrun Williksen-Bakker, "Vanua—A Symbol with Many Ramifications in Fijian Culture," *Ethnos* 55, nos. 3–4 (1990): 232–47, quotes at 235. A similar tradition exists in Hawai'i where a new mother's right to bring home her placenta was guaranteed by a 2006 law: HI Rev Stat § 321–30 (2023); see Celia T. Bardwell-Jones, "Placental Ethics: Addressing Colonial Legacies and Imagining Culturally Safe Responses to Health Care in Hawai'i," *The Pluralist* 13, no. 1 (2018): 97–114.
25. Williksen-Bakker, "Vanua," 235–36.
26. Jean-Pierre Labouisse, "Ethnobotany of Breadfruit in Vanuatu: Review and Prospects," *Ethnobiology Letters* 7, no. 1 (2016): 14–23. The sources that Labouisse cited for breadfruit diversity in the village of Ranon and the island of Mota are, respectively, C. Murray, "Varieties of Breadfruit, New Hebrides," *Journal of Polynesian Society* 3 (1894): 36; and R. H. Codrington, *The Melanesians: Studies in Their Anthropology and Folklore* (Oxford: Clarendon, 1891). On *Brassica oleracea*, see Lorenzo Maggioni, Roland von Bothmer, Gert Poulsen, and Ferdinando Branca, "Origin and Domestication of Cole Crops (*Brassica oleracea* L.): Linguistic and Literary Considerations," *Economic Botany* 64 (2010): 109–23, as well as Jeanne L. D. Osnas, "The Extraordinary Diversity of *Brassica oleracea*," *The Botanist in the Kitchen*, November 5, 2012, https://botanistinthekitchen.blog/2012/11/05/the-extraordinary-diversity-of-brassica-oleracea/. Many surveys have examined the diversity of breadfruit cultivars throughout the Pacific, and a few have done so in the Caribbean and other regions where breadfruit has been introduced. On Pacific breadfruit diversity, see for example, Diane Ragone, "Description of Pacific Island Breadfruit Cultivars," *Acta Horticulturae* 413 (1995): 93–98; Nyree J. C. Zerega, Diane Ragone, and Timothy J. Motley, "Complex Origins of Breadfruit (*Artocarpus altilis*, Moraceae): Implications for Human Migrations in Oceania," *American Journal of Botany* 91, no. 5 (2004): 760–66; A. Maxwell P. Jones, Susan J. Murch, Jim Wiseman, and Diane Ragone, "Morphological Diversity in Breadfruit (*Artocarpus*, Moraceae): Insights Into Domestication, Conservation, and Cultivar Identification," *Genetic Resources and Crop Evolution* 60 (2013): 175–92. On Caribbean breadfruit diversity, see for example, Oral O. Daley, Laura B. Roberts-Nkrumah, and Angela T. Alleyne, "Morphological Diversity of Breadfruit (*Artocarpus altilis* [Parkinson] Fosberg) in the Caribbean," *Scientia Horticulturae* 266 (2020): 109278; Laura B. Roberts-Nkrumah, *The Breadfruit Germplasm Collection at the University of the West Indies, St. Augustine Campus* (Mona, Jamaica: University of the West Indies Press, 2018).
27. Gloria Dickie, "The Last Tree Standing," *Modern Farmer*, November 5, 2018, https://modernfarmer.com/2018/11/the-last-tree-standing/; Carol Diane Ragone, "Collection, Establishment, and Evaluation of a Germplasm

Collection of Pacific Island Breadfruit," PhD diss. (University of Hawai'i-Mānoa, 1991).
28. "Svalbard Global Seed Vault," *Crop Trust*, https://www.croptrust.org/work/svalbard-global-seed-vault/.
29. Diana Noyce, "Charles Darwin, the Gourmet Traveler," *Gastronomica* 12, no. 2 (2012): 45–52, quotes throughout.

4. Awful and Lovely

1. J. M. Braga, *China Landfall: Jorge Alvares' Voyage to China, a Compilation of Some Relevant Material* (Macau: Imprensa Nacional, 1955). See especially 31–32 for a brief discussion of Alvares and Balboa's near-simultaneity. On Balboa's view of the Pacific, see Gerstle Mack, *The Land Divided: A History of the Panama Canal and Other Isthmian Canal Projects* (New York: Knopf, 1944), ch. 2; Simon Winchester, *Pacific: The Ocean of the Future* (London: Collins, 2015), 1–30; and Stefan Zweig, *Decisive Moments in History: Twelve Historical Miniatures*, trans. Lowell A. Bangerter (Riverside, CA: Ariadne Press, 1999), ch. 1.
2. Charles E. Nowell, "The Discovery of the Pacific: A Suggested Change of Approach," *Pacific Historical Review* 16, no. 1 (1947): 1–10, quote at 1.
3. Laurence Bergreen, *Over the Edge of the World: Magellan's Terrifying Circumnavigation of the Globe* (New York: Morrow, 2003), 1.
4. Gavan Daws, *Shoal of Time: A History of the Hawaiian Islands* (Honolulu: University of Hawai'i Press, 1974), 1; Woodrow Wilson, *A History of the American People* (New York: Harper, 1908), 1:32.
5. James Hoch, "Riding Backwards on a Train," in *A Parade of Hands: Poems* (Eugene, OR: Silverfish Review Press, 2003), 14.
6. Clements Markham, *The Voyages of Pedro Fernandez de Quiros, 1595 to 1606* (London: Hakluyt Society, 1904), 28.
7. William Dampier, *A new Voyage round the World* (London: Knapton, 1697), 296–97; Anna Neill, "Buccaneer Ethnography: Nature, Culture, and Nation in the Journals of William Dampier," *Eighteenth-Century Studies* 33, no. 2 (2000): 165–80. On Dampier, see Diana Preston and Michael Preston, *A Pirate of Exquisite Mind: Explorer, Naturalist, and Buccaneer—The Life of William Dampier* (New York: Walker and Company, 2004).
8. George Anson, *A Voyage Round the World in the Years MDCCXL, I, II, III, IV* (London: Knapton, 1749), 305, 310; Glyndwr Williams, *The Prize of All the Oceans: Commodore Anson's Daring Voyage and Triumphant Capture of the Spanish Treasure Galleon* (New York: Penguin, 2001); Georg Josef Kamel, "Plants of

Luzon Island: Historia stirpium insula Luzonis et reliquarum Philippinarum," in John Ray, *Historia plantarum* (London: Smith & Walford, 1704), 3:1–96.

9. Louis-Antoine de Bougainville, *The Pacific Journal of Louis-Antoine de Bougainville, 1767–1768*, ed. and trans. John Dunmore (London: Hakluyt Society, 2002), 238, edited for punctuation.

10. Bougainville, *The Pacific Journal*; Louis-Antoine de Bougainville, *Voyage autour du Monde, par la Frégate du Roi La Boudeuse, et la Flute l'Étoile* (Paris: Chez Sailland & Nyon, 1772). On Bougainville in the Pacific, see Vanessa Smith, *Intimate Strangers: Friendship, Exchange in Pacific Encounters* (Cambridge: Cambridge University Press, 2010); Andy Martin, "The Enlightenment in Paradise: Bougainville, Tahiti, and the Duty of Desire," *Eighteenth-Century Studies* 41, no. 2 (2008): 203–16; and Francis Leary, "'Tayo! Tayo!' in Nouvelle-Cythère," *Virginia Quarterly Review* 67, no. 4 (2003): 636–54.

11. Hesiod, *Theogony and Works and Days*, trans. Catherine M. Schlegel and Henry Weinfield (Ann Arbor: University of Michigan Press, 2006), 29; Philibert Commerson, "Sur la découverte de la nouvelle isle de Cythère ou Taïti," *Mercure de France* (November 1769), 197–207, reproduced in Bolton Glanvill Corney, *The Quest and Occupation of Tahiti by Emissaries of Spain During the Years 1772–1776* (London: Hakluyt Society, 1913), 2:461–66.

12. Genesis 3:19 (KJV).

13. Bougainville, *Voyage*, 2:45; Lord Byron, *The Island, or Christian and His Comrades* (London: Hunt, 1823), canto II, verse XI, lines 260–65.

14. James Cook, "Letter to John Walker, September 13, 1771," in *The Journals of Captain James Cook on His Voyages of Discovery*, vol. 1, *The Voyage of the Endeavour, 1768–1771*, ed. J. C. Beaglehole (Cambridge: Cambridge University Press, 1968), 507, edited for spelling.

15. On breadfruit's tolerance of various environmental conditions, see the systematic study conducted by Noa Kekuewa Lincoln, Alyssa Cho, Graham Dow, and Theodore Radovich, "Early Growth of Breadfruit in a Variety × Environment Trial," *Agronomy Journal* 111, no. 6 (2019): 3020–27.

16. Doug G. Sutton, *The Origins of the First New Zealanders* (Auckland, New Zealand: Auckland University Press, 1994); K. R. Howe, *The Quest for Origins: Who First Discovered and Settled the Pacific Islands?* (Honolulu: University of Hawai'i Press, 2003); Janet M. Wilmshurst, Atholl J. Anderson, Thomas F. G. Higham, and Trevor H. Worthy, "Dating the Late Prehistoric Dispersal of Pacific Islanders to New Zealand Using the Commensal Pacific Rat," *Proceedings of the National Academy of Sciences of the United States* 105, no. 2 (2008): 7676–80; Anne Salmond, *Aphrodite's Island: The European Discovery of Tahiti* (Berkeley: University of California Press, 2009); Fiona Petchey and Magdalena M. E. Schmid, "Vital Evidence: Change in the Marine ^{14}C Reservoir Around New Zealand

(Aotearoa) and Implications for the Timing of Polynesian Settlement," *Scientific Reports* 10 (2020): 14266.
17. Charles Hursthouse, *New Zealand, or Zealandia, the Britain of the South* (London: Edward Stanford, 1857), 101. Sociologist Max Weber developed the concept of a "Protestant work ethic" in *The Protestant Ethic and the Spirit of Capitalism*, trans. Talcott Parsons (New York: Scribner, 1930), originally published in German in 1905; Edna Dean Proctor, *The Glory of Toil and Other Poems* (Boston: Houghton Mifflin, 1916), 1.
18. Syed Hussein Alatas, *The Myth of the Lazy Native: A Study of the Image of the Malays, Filipinos and Javanese from the 16th to the 20th Century and Its Function in the Ideology of Colonial Capitalism* (London: Frank Cass, 1977). See also Edward Said, *Orientalism* (New York: Pantheon, 1978).
19. Alexander von Humboldt and Aimé Bonpland, *Personal Narrative of Travels to the Equinoctial Regions of the New Continent During the Years 1799–1804*, trans. Helen Maria Williams (London: Longman, Hurst, Rees, Orme and Brown, 1818), 3:14–15; See also Andrea Wulf, *The Invention of Nature: Alexander von Humboldt's New World* (New York: Knopf, 2015).
20. Philip H. Young, "The Cypriot Aphrodite Cult: Paphos, Rantidi, and Saint Barnabas," *Journal of Near Eastern Studies* 64, no. 1 (2005): 23–44, quotes at 23.
21. Hursthouse, *New Zealand*, 17, 102–4.
22. Basil Keane, "Kurī—Polynesian Dogs," November 24, 2008, http://www.TeAra.govt.nz/en/kuri-polynesian-dogs, and Gerard Hutching, "Sharks and Rays—Māori and sharks," June 12, 2006, http://www.TeAra.govt.nz/en/sharks-and-rays, both in *Te Ara: The Encyclopedia of New Zealand*; Paul Moon, *This Horrid Practice: The Myth and Reality of Traditional Maori Cannibalism* (Auckland, New Zealand: Penguin Random House New Zealand Limited, 2008).
23. On American and European castaways who integrated, to varying degrees, into Pacific Islander societies, see I. C. Campbell, *"Gone Native" in Polynesia: Captivity Narratives and Experiences from the South Pacific* (Westport, CT: Greenwood, 1998).
24. Paul Gauguin, *Te raau rahi (The Big Tree)*, Art Institute Chicago, https://www.artic.edu/artworks/111062/te-raau-rahi-the-big-tree; Paul Gauguin, *Ia Orana Maria (Hail Mary)*, Metropolitan Museum of Art, https://www.metmuseum.org/art/collection/search/438821; Robert Louis Stevenson, *Ballads* (New York: Charles Scribner's Sons, 1890), 21, 37.
25. J. M. Barrie, *Peter Pan and Wendy* (New York: Charles Scribner's Sons, 1921), 83. David Park Williams, "Hook and Ahab: Barrie's Strange Satire on Melville," *PMLA* 80, no. 5 (1965): 483–88, quote at 483.
26. Herman Melville, *Typee: A Peep at Polynesian Life* (London: John Murray, 1846), 79.

27. Hershel Parker, *Herman Melville: A Biography* (Baltimore: Johns Hopkins University Press, 1996), 1:184; Herman Melville, *Moby-Dick; or, The Whale* (New York: Harper, 1851), 123; Melville, *Typee*, 20.
28. Melville, *Moby-Dick*, 96–97.
29. Herman Melville, *Typee: A Romance of the South Seas*, ed. Sterling Andrus Leonard (New York: Harcourt, Brace, and Howe, 1920), 126–28, edited for spelling.
30. Jules Verne, *Twenty Thousand Leagues Under the Sea* (New York: Butler Brothers, 1887), 129–32.
31. Coleridge's suggestion comes by way of a letter written by the poet Peter Southey and cited in Peter J. Kitson, "Sustaining the Romantic and Racial Self: Eating People in the 'South Seas,'" in *Cultures of Taste/Theories of Appetite: Eating Romanticism*, ed. Timothy Morton (New York: Palgrave Macmillan, 2004), 77–96, quote at 92, edited for punctuation; James Morrison, *The Journal of James Morrison, Boatswain's Mate on the* Bounty (London: The Golden Cockerel Press, 1935), 201–2.
32. Teva Beguet, email to author, April 19, 2022.
33. Morgain Halberg, "Peek Inside Barack Obama's Exclusive Island Writing Retreat," *Observer*, March 28, 2017, https://observer.com/2017/03/barack-obama-the-brando-resort-writing-memoir/. On Teti'aroa and the Teti'aroa Society more generally, see Hampton Sides, "We Had Marlon Brando's Island Utopia to Ourselves," *Outside*, March 1, 2021, https://www.outsideonline.com/adventure-travel/essays/the-brando-resort/.
34. M. H. Sachet and F. R. Fosberg, *An Ecological Reconnaissance of Tetiroa Atoll, Society Islands* (Washington, DC: Smithsonian Institution, 1983).
35. Marlon Brando, *Brando: Songs My Mother Taught Me* (New York: Random House, 1994), 269.
36. Julian Sancton, "Last Tango on Brando Island," *Maxim*, December 11, 2008, https://www.maxim.com/entertainment/last-tango-brando-island/.

5. The Wondrous Food of the Land

1. J. C. Beaglehole, ed. *The Journals of Captain James Cook on his Voyages of Discovery*, vol. 3, pt. 1: *The Voyage of the Resolution and Discovery 1776–1780* (London: Routledge, 2017), 279.
2. John Reinhold Forster, *Observations made during a Voyage round the World* (London: Robinson, 1778), 512; Litton Forbes, "The Navigator Islands," *Proceedings of the Royal Geographical Society of London* 21, no. 2 (1876–1877): 140–48, quote at 141; Andía y Varela cited in Bolton Glanvill Corney, *The Quest and Occupation*

of *Tahiti by Emissaries of Spain* (London: Hakluyt Society, 1914), 2:284, 286; David Lewis, *We, the Navigators: The Ancient Art of Landfinding in the Pacific* (Honolulu: University of Hawai'i Press, 1972), 19.

3. Andrew Sharp, *Ancient Voyagers in the Pacific* (London: Penguin, 1957); Andrew Sharp, "Polynesian Navigation to Distant Islands," *Journal of the Polynesian Society* 70, no. 2 (1961): 219–26. For a retrospective analysis seemingly intended to soften Sharp's position, see K. R. Howe, "Voyagers and Navigators: The Sharp-Lewis Debate," in *Texts and Contexts: Reflections in Pacific Islands Historiography*, ed. Doug Munro and Brij V. Lal (Honolulu: University of Hawai'i Press, 2006), 65–75.

4. P. A. M. Dirac, *The Principles of Quantum Mechanics* (Oxford: Oxford University Press, 1930), v.

5. David Lewis, "Polynesian and Micronesian Navigation Techniques," *Journal of Navigation* 23, no. 4 (1970): 432–47, quotes at 433.

6. Ben Finney, *Voyage of Rediscovery: A Cultural Odyssey through Polynesia* (Berkeley: University of California Press, 1994), 34.

7. Finney, *Voyage of Rediscovery*, 42.

8. On the inclusion of ʻulu among Hōkūleʻa's provisions, I relied upon personal communication from Nainoa Thompson of the Polynesian Voyaging Society, April 16, 2024.

9. Francisco García, *Vida y Martyrio de el Venerable Padre Diego Luis de Sanvitores de la Compaña de Iesus, Primer Apostol de las Islas Marianas y Sucessos de estas Islas desde el Año de Mil Seiscientos y Sesenta y Ocho, asta el de Mil Seiscientos y Ochenta y Uno* (Madrid: Ivan García Infanzón, 1683), 197.

10. Roger Haden, *Food Culture in the Pacific Islands* (Santa Barbara, CA: ABC-CLIO, 2009), 17.

11. Haden, *Food Culture*, 18, 20–21.

12. Haden, *Food Culture*, 21; Jane Parry, "Pacific Islanders Pay Heavy Price for Abandoning Traditional Diet," *Bulletin of the World Health Organization* 88 (2010): 484–85.

13. James Cook, *The Journals of Captain James Cook on his Voyages of Discovery*, vol. 2, *The Voyage of the Resolution and Adventure 1772–1775*, ed. J. C. Beaglehole (London: Routledge, 2017), 175, edited for spelling.

14. On the kaluʻulu, see Noa Lincoln and Thegn Ladefoged, "Agroecology of Pre-contact Hawaiian Dryland Farming: The Spatial Extent, Yield and Social Impact of Hawaiian Breadfruit Groves in Kona, Hawaiʻi," *Journal of Archaeological Science* 49 (2014): 192–202. On the replacement of breadfruit with coffee in the kaluʻulu, see the comments made by Noa Kekuewa Lincoln on the episode of the *Gastropod* podcast titled, "The Fruit That Could Save the World," https://gastropod.com/transcript-the-fruit-that-could-save-the-world/.

15. "Māla Kaluʻulu," Hawaiʻi ʻUlu Cooperative, https://eatbreadfruit.com/blogs/meet-our-farmers/mala-kaluulu.
16. Bill Goodwin, *Frommer's Tahiti and French Polynesia* (New York: Wiley, 2010), 99.
17. The Samoan term *ulu* is a cognate of the Hawaiian *ʻulu* but does not include the *ʻokina* (ʻ).
18. Sam Choy and Gay Wong, *Sam Choy's ʻUlu Cookbook: Hawaiʻi's Breadfruit Recipes* (Honolulu: Mutual Publishing, 2022), 3. On "Sam Choy's in the Kitchen," see Catherine Toth Fox, "Sam Choy Wants to Cook in Your Kitchen," *Honolulu*, March 16, 2016, https://www.honolulumagazine.com/sam-choy-wants-to-cook-in-your-kitchen/.
19. Gordon Ramsay "Hana Coast," *Uncharted*, National Geographic Channel, August 11, 2019. On axis deer, see Elena C. Rubino and Christopher K Williams, "Exploring Public Support for Large-Scale Commercial Axis Deer Harvests in Maui, Hawaii," *Sustainability* 14, no. 3 (2022): 1837.
20. On *haoles* in Hawaiʻi, see Judy Rohrer, *Haoles in Hawaiʻi* (Honolulu: University of Hawaiʻi Press, 2010).
21. "About the Hawaiʻi ʻUlu Cooperative," Hawaiʻi ʻUlu Cooperative, https://eatbreadfruit.com/pages/aboutus.
22. "Certified Copy of Will and Codicil, Supreme Court of the Territory of Hawaii, October Term 1942," King Lunalilo Trust, https://www.lunalilo.org/images/KingLunalilo_will_and_codicil.pdf.
23. "History of King Lunalilo Trust and Lunalilo Home," King Lunalilo Trust, https://www.lunalilo.org/about/history-of-lunalilo-home.
24. On the roles of food and music in helping dementia patients to recover memories, see Peggy Orenstein, "The Power of Food for People with Dementia," *New Yorker*, September 5, 2023, https://www.newyorker.com/culture/personal-history/the-power-of-food-for-people-with-dementia; Bill Mossman, "The Right Person for Lunalilo," *MidWeek*, April 29, 2020, https://www.midweek.com/diane-paloma-lunalilo-home/.
25. Matthew K. Loke and PingSun Leung, "Hawaiʻi's Food Consumption and Supply Sources: Benchmark Estimates and Measurement Issues," *Agricultural and Food Economics* 1 (2013): 10.
26. On the Thirty Meter Telescope controversy, see Marie Alohalani Brown, "Mauna Kea: Hoʻomana Hawaiʻi and Protecting the Sacred," *Journal for the Study of Religion, Nature and Culture* 10, no. 2 (2016): 150–69, and Iokepa Casumbal-Salazar, "A Fictive Kinship: Making 'Modernity,' 'Ancient Hawaiians,' and the Telescopes of Mauna Kea," *Native American and Indigenous Studies* 4, no. 2 (2017): 1–30.
27. Israel "IZ" Kamakawiwoʻole, "Hawaiʻi '78," on *Facing Future* (Mountain Apple Company/Big Boy Records, 1993).

28. Michael Pollan, *The Omnivore's Dilemma: A Natural History of Four Meals* (New York: Penguin, 2006), 259. Wendell Berry's quote appeared originally in his 1989 essay "The Pleasures of Eating," which can be found in several compilations including *Bringing It to the Table: On Farming and Food* (Berkeley, CA: Counterpoint, 2009), 282.
29. The text of Queen Liliʻuokalani's statement, by which she yielded control of Hawaiʻi (under duress) to the United States, was printed in a letter addressed to Sanford B. Dole, president of the Provisional Government of Hawaii, which served as a transition between the overthrown kingdom and the statuses of republic and U.S. territory that would follow. The text and a facsimile of the original document can be found at the website of the library of the University of Hawaiʻi at Mānoa, https://libweb.hawaii.edu/digicoll/annexation/protest/liliu2.php.
30. Many recordings of "Aloha ʻoe" exist, including versions sung by Bing Crosby and Elvis Presley. The original lyrics and music were published as "Aloha Oe, Song and Chorus, Composed by Her Royal Highness Princess Liliuokalani of Honolulu, Oahu, H.I." (San Francisco: The Pacific Music Company, 1890).
31. On the political implications of Prendergast's song see Eleanor C. Nordyke and Martha H. Noyes, " 'Kaulana Nā Pua': A Voice for Sovereignty," *Hawaiian Journal of History* 27 (1993), 27–42. The version of the lyrics I use here were quoted and translated by Noenoe Silva in Anne Keala Kelly, "Kūkākūkā (to Discuss)," *Honolulu Weekly*, January 8, 2003.
32. Exodus 16:20–21 (KJV); Luke 11:3 (KJV); Noa Kekuewa Lincoln, Diane Ragone, Nyree J. C. Zerega, Laura B. Roberts-Nkrumah, Mark Merlin, and A. Maxwell P. Jones, "Grow Us Our Daily Bread: A Review of Breadfruit Cultivation in Traditional and Contemporary Systems," *Horticultural Reviews* 46 (2019): 299–384.

6. A Dish de Résistance

1. Hannah Rachel Cole, "Breadfruit in the Wake: Imagining Vegetal Mutiny in Derek Walcott's 'The Bounty,' " *Latin American Literary Review* 48, no. 96 (2021): 35–39.
2. Derek Walcott, "A Sea Chantey," *The Poetry of Derek Walcott, 1948–2013*, ed. Glyn Maxwell (New York: Farrar, Straus and Giroux, 2014), 51. On Caribbean cuisine, see Lynn Marie Houston, *Food Culture in the Caribbean* (London: Greenwood, 2005) and Barry W. Higman, "Cookbooks and Caribbean Cultural Identity: An English-Language Hors d'Oeuvre," *New West Indian Guide/Nieuwe West-Indische Gids* 72, nos. 1–2 (1998): 77–95. A Caribbean cookbook that I

recommend for its diversity of islands represented is Jessica B. Harris, *Sky Juice and Flying Fish: Traditional Caribbean Cuisine* (New York: Simon & Schuster, 1991).

3. V. S. Naipaul, *The Middle Passage: Impressions of Five Societies—British, French, and Dutch—in the West Indies and South America* (London: Readers Union/Andre Deutsch, 1963), 182; Jamaica Kincaid, *Annie John* (New York: New American Library, 1986), 83; Jamaica Kincaid, *My Garden (Book)* (New York: Farrar, Straus and Giroux, 1999), 135; Michel Erman, "What Is a National Dish?," *Médium* 28, no. 3 (2011): 31–43, quote at 31.

4. David Lowenthal, *West Indian Societies* (London: Oxford University Press, 1972), 68.

5. The celebrity in question was the actor Hilary Swank, who was fined NZ$200 in 2005 (about US$140 at the time) for failing to declare both an apple and an orange in her bag. "Hollywood Actress Fined for Bringing Fruit Into NZ," *New Zealand Herald*, https://www.nzherald.co.nz/nz/hollywood-actress-fined-for-bringing-fruit-into-nz/V6N2LZWBDOXJXTRX2GKGWVQYPU/. On economic botany see Richard Harry Drayton, *Nature's Government: Science, Imperial Britain, and the "Improvement" of the World* (New Haven, CT: Yale University Press, 2000), and James E. McClellan, *Colonialism and Science: Saint Domingue in the Old Regime* (Chicago: University of Chicago Press, 2010), ch. 9.

6. Alfred W. Crosby Jr., *The Columbian Exchange: Biological and Cultural Consequences of 1492* (Westport, CT: Greenwood Press, 1972); Londa Schiebinger, *Plants and Empire: Colonial Bioprospecting in the Atlantic World* (Cambridge, MA: Harvard University Press, 2009), 7.

7. For one telling of the history of the introduction of vervet monkeys to the Caribbean, see Jean-Baptiste Labat, *The Memoirs of Père Labat, 1693–1705*, trans. John Eaden (London: Cass, 1970), 60–61; on their present status, see Kerry M. Dore, "Vervets in the Caribbean," in *The International Encyclopedia of Primatology*, ed. Agustín Fuentes (Hoboken, NJ: Wiley, 2017), 1425–28, as well as other work by Dore.

8. Richard H. Grove, *Green Imperialism: Colonial Expansion, Tropical Island Edens and the Origins of Environmentalism, 1600–1860* (Cambridge: Cambridge University Press, 1995), 339; Toby Musgrave, *The Multifarious Mr. Banks: From Botany Bay to Kew, the Natural Historian Who Shaped the World* (New Haven, CT: Yale University Press, 2020), 190; Douglas Oliver, *Return to Tahiti: Bligh's Second Breadfruit Voyage* (Honolulu: University of Hawai'i Press, 1988).

9. Drayton, *Nature's Government*, 12; John Keay, *India: A History* (New York: Grove Press, 2010); Banks quoted in Grove, *Green Imperialism*, 338.

10. Jamaica Kincaid, "The Breadfruit," *Harvard Advocate*, Winter 2015, https://theharvardadvocate.com/content/the-breadfruit.

11. Published accounts of the *Bounty* voyage abound. I have relied mainly on the following sources: William Bligh, *A Narrative of the Mutiny on board His Majesty's Ship Bounty* (London: Nicol, 1790); Richard Hough, *Captain Bligh & Mr. Christian: the Men and the Mutiny* (London: Hutchinson, 1972); William Bligh, *A Voyage to the South Sea, Undertaken by Command of His Majesty, for the Purpose of Conveying the Breadfruit Tree to the West Indies, in His Majesty's Ship the Bounty* (London: Nicol, 1792); and Jennifer Newell, *Trading Nature: Tahitians, Europeans, & Ecological Exchange* (Honolulu: University of Hawai'i Press, 2010).
12. Bligh, *A Narrative of the Mutiny*, 14; Hough, *Captain Bligh & Mr. Christian*, 123. On Mauatua and her relationship with Fletcher Christian, see *Mrs. Christian: Bounty Mutineer* (London: Hendon, 2019), by Glynn Christian, a direct descendant of their union.
13. Hough, *Captain Bligh & Mr. Christian*, 132.
14. Hough, *Captain Bligh & Mr. Christian*, 134, 143–145.
15. Bligh, *A Narrative of the Mutiny*, 2–3.
16. Hough, *Captain Bligh & Mr. Christian*, 157.
17. Hough, *Captain Bligh & Mr. Christian*, 159.
18. Newell, *Trading Nature*, 158; Bligh, *A Narrative of the Mutiny*, 11. The original quote used the now-obsolete British name for Tahiti, "Otaheite."
19. Bligh, *A Narrative of the Mutiny*, 8, 138–39; George Tobin, *Journal of Lieutenant George Tobin on HMS* Providence *1791–1793*, ML A562, CY 1421, Mitchell Library, State Library of New South Wales, Sydney, http://acms.sl.nsw.gov.au/_transcript/2011/D04424/a1220.htm, 301.
20. On Bligh's tomb, see Christopher Woodward, "Captain Bligh's Tomb," *Australian Garden History* 27, no. 4 (2017): 18–20.
21. "Pitcairn Island," written by Julien Arnold and Scott Peters, performed by Captain Tractor, https://captaintractor.com.
22. On the pre-European history of Pitcairn Island, see Guillaume Molle and Aymeric Hermann, "Pitcairn Before the Mutineers: Revisiting the Isolation of a Polynesian Island," in *The Bounty from the Beach: Cross-Cultural and Cross-Disciplinary Essays*, ed. Sylvie Largeaud-Ortega (Canberra: Australian National University Press, 2018), 67–94. On the post-mutiny history of Pitcairn Island, I have relied mainly upon the website, "History of Pitcairn Island," hosted by the Pitcairn Islands Study Center at Pacific Union College, https://library.puc.edu/pitcairn/pitcairn/history.shtml.
23. On the Pitcairn Island sexual abuse trials, see Dawn Oliver, ed., *Justice, Legality and the Rule of Law: Lessons from the Pitcairn Prosecutions* (Oxford: Oxford University Press, 2009) and Kathy Marks, *Lost Paradise: From Mutiny on the Bounty to a Modern-Day Legacy of Sexual Mayhem, the Dark Secrets of Pitcairn Island Revealed* (New York: Simon & Schuster, 2009).

24. The "hunting" of breadfruit with rifles and shotguns has been recorded in multiple accounts of travel to the island, including Ben Fogle, *The Teatime Islands: Journeys to Britain's Faraway Outposts* (London: Michael Joseph, 2003), and Marks, *Lost Paradise*.

7. An Idea Whose Time Had Come

1. Hugo wrote in *Histoire d'un crime: Déposition d'un témoin* (Paris: Lévy, 1877), 300, "On résiste à l'invasion des armées; on ne résiste pas à l'invasion des idées," which can be translated, as it was by Huntington Smith, in *History of a Crime* (New York: Crowell, 1888), 237, "an invasion of armies can be resisted, but there is no resistance to an invasion of ideas." Aimard, however, was much closer to the exact phrase in circulation today when he wrote in *Les francs tireurs* (Paris: Amyot, 1861), 68, "dans toute question humaine, il y a quelque chose de plus puissant que la force brutale des baïonnettes: c'est l'idée dont le temps est venu et l'heure est sonnée," which was translated by Lascelles Wraxall in *The Freebooters: A Story of the Texan War* (London: Ward and Lock, 1861), 52, as "in every human question, there is something more powerful than the brute force of bayonets: it is the idea whose time has come and hour struck."
2. Paul Fussell Jr., "Patrick Brydone: The Eighteenth-Century Traveler as Representative Man," *Bulletin of the New York Public Library* 66, no. 6 (1962): 349–63, quotes at 350.
3. John T. Schlebecker, "Farming at the Time of the Constitution," in *Our American Land: 1987 Yearbook of Agriculture*, ed. William Whyte (Washington, DC: U.S. Department of Agriculture, 1987), 26–30, quote at 26, edited for spelling.
4. Vetch quoted in Cecil Headlam, ed., *Calendar of State Papers, Colonial Series, America and West Indies, June, 1708–1709* (London: His Majesty's Stationary Office, 1922), 47.
5. Agnes M. Whitson, "The Outlook of the Continental American Colonies on the British West Indies, 1760–1775," *Political Science Quarterly* 45, no. 1 (1930): 56–86; Matthew Mulcahy, *Hubs of Empire: The Southeastern Lowcountry and British Caribbean* (Baltimore: Johns Hopkins University Press, 2014), 43; Mark Kurlansky, *Cod: A Biography of the Fish That Changed the World* (New York: Penguin, 1998).
6. Kurlansky, *Cod*, 80–81.
7. Simon Taylor, writing in 1784, quoted in Jennifer Newell, *Trading Nature: Tahitians, Europeans, and Ecological Exchange* (Honolulu: University of Hawai'i Press, 2010), 147.
8. John Hamilton cited in Ron Chernow, *Alexander Hamilton* (New York: Penguin, 2005), 29.

9. Alexander Hamilton and the Continental Congress both cited in Richard B. Sheridan, "The Crisis of Slave Subsistence in the British West Indies During and After the American Revolution," *William and Mary Quarterly* 33, no. 4 (1976): 615–41, quotes at 616–617, edited for punctuation.
10. John Hawkesworth, *An Account of the Voyages Undertaken by the Order of His Present Majesty for Making Discoveries in the Southern Hemisphere* (London: Strahan and Cadell, 1773); Newell, *Trading Nature*.
11. Hawkesworth, *An Account of the Voyages*, 80–81.
12. Hawkesworth, *An Account of the Voyages*, 197, edited for punctuation.
13. Emma C. Spary and Paul White, "Food of Paradise: Tahitian Breadfruit and the Autocritique of European Consumption," *Endeavour* 28, no. 2 (2004): 75–80, quote at 75; Elizabeth M. DeLoughrey, "Globalizing the Routes of Breadfruit and Other Bounties," *Journal of Colonialism and Colonial History* 8, no. 3 (2007), para. 45; Greg Dening, *Mr. Bligh's Bad Language: Passion, Power and Theater on the Bounty* (Cambridge: Cambridge University Press, 1992), 11.
14. Michael Morrissey, *Breadfruit Stories: A Tree's Journey from Tahiti to the West Indies, 1760s to 1840s* (Winnetka, IL: Trees That Feed Foundation, 2021), 29.
15. David Mackay, "Banks, Bligh and Breadfruit," *New Zealand Journal of History* 8, no. 1 (1974): 61–77, quote at 61.
16. MacKay, "Banks, Bligh, and Breadfruit," 63; Rebecca Earle, "Food, Colonialism and the Quantum of Happiness," *History Workshop Journal* 84 (2017): 170–93, quote at 176.
17. Toby Musgrave, *The Multifarious Mr. Banks: From Botany Bay to Kew, The Natural Historian Who Shaped the World* (New Haven, CT: Yale University Press, 2020), 138; Jordan Goodman, *Planting the World: Joseph Banks and His Collectors, an Adventurous History of Botany* (London: William Collins, 2020), 123.
18. Charles Bucke, *On the Beauties, Harmonies, and Sublimities of Nature: With Occasional Remarks on the Laws, Customs, Manners, and Opinions of Various Nations* (London: G. & W. B. Whittaker, 1823), 3:321; Richard A. Howard, "A History of the Botanic Garden of St. Vincent, British West Indies," *Geographical Review* 44, no. 3 (1954): 381–93, quote at 383; William Coxe, *An Historical Tour in Monmouthshire* (London: Cadell and Davies, 1801), 394.
19. Valentine Morris to Joseph Banks, April 13, 1772, British Library (Add. Ms 33977.18); Ivor Waters, *The Unfortunate Valentine Morris* (Chepstow: Chepstow Society, 1964).
20. Julia Bruce, "Banks and Breadfruit," *RSA Journal* 141, no. 5444 (1993): 818–19.
21. Juliane Braun, "Bioprospecting Breadfruit: Imperial Botany, Transoceanic Relations, and the Politics of Translation," *Early American Literature* 54, no. 3 (2019): 643–71, quote at 664n2. Braun provided the translation quoted here; all other translations of excerpts from Voltaire are my own. Voltaire, "Arbre à

pain," in *Questions sur l'encyclopédie par des amateurs* (Geneva: Cramer, 1770), 2:104–7, quote at 104.
22. James Hanrahan, "A Significant Enlightenment Text Fallen Between the Cracks of Editorial History," *Eighteenth-Century Studies* 45, no. 1 (2011): 157–60, quote at 157; "Questions sur l'Encyclopédie (I–VIII)," *Voltaire Foundation*, http://www.voltaire.ox.ac.uk/publication/questions-sur-lencyclop%c3%a9die-i-viii/.
23. Voltaire, "Arbre à pain," 105; Antony Wild, *Coffee: A Dark History* (New York: Norton, 2004); William H. Ukers, *All About Coffee* (New York: The Tea and Coffee Trade Journal Company, 1922), 6.
24. Voltaire, "Arbre à pain," 105–6.
25. Waters, *The Unfortunate Valentine Morris*; Ivor Waters, *Chepstow Packets* (Chepstow: Moss Rose Press, 1983); Philip Gosse, *Dr. Viper: The Querulous Life of Philip Thicknesse* (London: Cassell, 1952), 282; Voltaire, *The Works of M. de Voltaire*, trans. T. Smollett, T. Francklin, et al. (London: Newberry, 1761), 1:iii–iv; Philip Thicknesse, *Memoirs and Anecdotes of Philip Thicknesse: Late Lieutenant Governor of Land Guard Fort, and, Unfortunately, Father to George Touchet, Baron Audley* (London: s.n, 1791), 3:157; Ifan Kyrle Fletcher, "What Thicknesse Read," *Book-Collector's Quarterly* 16 (1934): 49–57, quote at 50.
26. Morris to Banks, April 13, 1772.

8. No Poem About Breadfruit

1. "From Alexander Anderson," in *The Indian and Pacific Correspondence of Sir Joseph Banks, 1768–1820*, ed. Neil Chambers (London: Pickering & Chatto, 2011), 4:371–73, quote at 371.
2. John H. Parry, "Plantation and Provision Ground: An Historical Sketch of the Introduction of Food Crops Into Jamaica," *Revista de Historia de América* 39 (1955): 1–20, quote at 19; Jill H. Casid, *Sowing Empire: Landscape and Colonization* (Minneapolis: University of Minnesota Press, 2005), 23; David Watts, *The West Indies: Patterns of Development, Culture and Environmental Change Since 1492* (Cambridge: Cambridge University Press, 1990), 505.
3. David Lowenthal, *West Indian Societies* (London: Oxford University Press, 1972), 68.
4. Gordon Rohlehr, *Calypso & Society in Pre-Independence Trinidad* (Port of Spain, Trinidad: Gordon Rohlehr, 1990), 221; Édouard Glissant, *Poetics of Relation*, trans. Betsy Wing (Ann Arbor: University of Michigan Press, 1997), 148, originally published in French in 1990.
5. Jamaica Kincaid, *My Garden (Book)* (New York: Farrar, Straus and Giroux, 1999), 135; Jamaica Kincaid, *Annie John* (New York: New American Library, 1986), 83–84.

6. Wayne Kaumualii Westlake, "Breadfruit," in *Westlake: Poems by Wayne Kaumualii Westlake (1947–1984)*, ed. Mei-Li M. Siy and Richard Hamasaki (Honolulu: University of Hawai'i Press, 2009), 192; Diane Ako, "Breadfruit Thrown Through Bus Window, Breaks Woman's Nose," *KITV Island News*, December 31, 2017, https://www.kitv.com/story/37168325/breadfruit-thrown-through-bus-window-breaks-womans-nose.
7. Mathieu Morin, "The West Indian Diet: Thomas Dancer, Breadfruit and Fever Epidemics in Eighteenth Century Jamaica," MA thesis, (Concordia University, 2018), 85 (emphasis added).
8. Robert A. Hill, ed. *The Marcus Garvey and Universal Negro Improvement Association Papers* (Berkeley: University of California Press, 1990), 7:791; Bob Marley and the Wailers, "Redemption Song," on *Uprising* (Tuff Gong/Island Records, 1980).
9. Hilary McD. Beckles, *How Britain Underdeveloped the Caribbean: A Reparation Response to Europe's Legacy of Plunder and Poverty* (Mona, Jamaica: University of the West Indies Press, 2022). The title of Beckles's book consciously draws upon that of a classic of anticolonial literature, Walter Rodney's *How Europe Underdeveloped Africa* (London: Bogle-L'Ouverture, 1972). Beckles's book was published on the fiftieth anniversary of the publication of Rodney's.
10. Demetrius L. Eudell, *The Political Languages of Emancipation in the British Caribbean and the U.S. South* (Chapel Hill: University of North Carolina Press, 2002), 131.
11. Gerald Oster and Selmaree Oster, "The Great Breadfruit Scheme," *National History* 94, no. 3 (1985): 35–41, quote at 35.
12. The Derek Walcott interview quoted here is unpublished but was attributed to Carrol B. Fleming and cited in Erika J. Smilowitz, "Fruits of the Soil: Botanical Metaphors in Caribbean Literature," *World Literature Written in English* 30, no. 1 (1990): 29–36, quotes at 30.
13. On colorism, or discrimination based upon skin tone, see Ronald E. Hall, ed., *Interdisciplinary Perspectives on Colorism: Beyond Black and White* (New York: Taylor & Francis, 2023); Kathy Russell, Midge Wilson, and Ronald E. Hall, eds., *The Color Complex: The Politics of Skin Color Among African Americans* (New York: Anchor, 1993); and Sarah L. Webb, "Colorism Healing," https://thechcoffeehouse.colorismhealing.com/.
14. Joshua Jelly-Schapiro, "Introduction," in Patrick Leigh Fermor, *The Traveller's Tree: A Journey Through the Caribbean Islands* (London: John Murray, 1950), viii.
15. Fermor, *The Traveller's Tree*, 88.
16. Andrea Stuart, *Sugar in the Blood: A Family's Story of Slavery and Empire* (New York: Knopf, 2013), 309.
17. James Mitchell, *Beyond the Islands: An Autobiography* (London: Macmillan Caribbean, 2006), 444; Richard A. Byron-Cox, "James Mitchell: A Myth of Great

Premiership," *One News St. Vincent*, August 16, 2020, https://onenewsstvincent.com/2020/08/16/james-mitchell-a-myth-of-great-premiership/.

18. Benjamin Franklin, *Poor Richard's Almanack* (New York: Century, 1898), 118; Matthew 6:11 (KJV).

19. Genesis 1:29 (KJV); Jean-Jacques Rousseau, *The Social Contract & Discourses*, trans. G. D. H. Cole (London: Dent, 1920), 207.

20. Daniel Quinn, *Ishmael: An Adventure of the Mind and Spirit* (New York: Bantam, 1992); Baylen J. Linneken, *Biting the Hands that Feed Us: How Fewer, Smarter Laws Would Make Our Food System More Sustainable* (Washington, DC: Island Press, 2016); Baylen J. Linneken, "Food Law Gone Wild: The Law of Foraging," *Fordham Urban Law Journal* 45, no. 4 (2018): 995–1050; Jared Diamond, "The Worst Mistake in the History of the Human Race," *Discover* 8, no. 5 (1987): 64–66.

21. William Dampier, *A new Voyage round the World* (London: Knapton, 1697), 296–97; Vanessa Smith, "Give Us Our Daily Breadfruit: Bread Substitution in the Pacific in the Eighteenth Century," *Studies in Eighteenth Century Culture* 35 (2006): 53–75, quote at 55. The question, "Who is the Tolstoy of the Zulus . . . the Proust of the Papuans?" is said to have been asked derisively by the writer Saul Bellow. In *Between the World and Me* (New York: Spiegel & Grau, 2015), Ta-Nehisi Coates cited the profoundly simple response, "Tolstoy is the Tolstoy of the Zulus," from journalist Ralph Wiley's *Dark Witness: When Black People Should Be Sacrificed (Again)* (New York: Ballantine, 1996). Bellow denied having ever asked the question or even having expressed the sentiment it conveys. The quote is found at 56 in Coates, 31–32 in Wiley. Bellow's response is found in "Op-Ed: Papuans and Zulus," *New York Times*, March 10, 1994, https://www.nytimes.com/books/00/04/23/specials/bellow-papuans.html.

22. James Boswell, *The Life of Samuel Johnson, LL.D.* (London: Henry Baldwin, 1791), 1:414–15, edited for punctuation and spelling.

23. London Missionary Society, *A Missionary Voyage to the Southern Pacific Ocean, Performed in the Years 1796, 1797, 1798, in the Ship* Duff, *Commanded by Captain James Wilson* (London: Gillet, 1799), 81; Charles Dolman, "The Wesleyan Missions at Otaheite," *Dolman's Magazine* 8, no. 44 (October 1848), 197–214, quote at 209, edited for spelling.

24. Bob Blaisdell, ed., *The Wit and Wisdom of Mark Twain* (Mineola, NY: Dover, 2013), 99; Sanford Niles, *The Elementary Geography* (Indianapolis: Indiana School Book Company, 1889), 17.

25. The identification of "Western, educated, industrialized, rich, and democratic" societies by the acronym WEIRD is attributed to anthropologist Joseph Henrich, who coined the term in a paper with coauthors Steven J. Heine and Ara Norenzayan, "The Weirdest People in the World?," *Behavioral and Brain Sciences* 33, no. 2–3 (2010): 61–83, and popularized it in his book, *The WEIRDest People in the World: How the West Became Psychologically Peculiar and Particularly*

Prosperous (New York: Farrar, Straus and Giroux, 2020); Thomas Hobbes, *Leviathan, or The Matter, Forme, & Power of a Common-Wealth Ecclesiastical and Civil* (London: Crooke, 1651), 62, 103.

26. Marshall Sahlins, "Notes on 'The Original Affluent Society,'" in *Man the Hunter*, ed. Richard B. Lee and Irven DeVore (Hawthorne, NY: Wenner-Gren, 1968), 85–89. Quinn, *Ishmael*, 239.
27. Milton Friedman, *There's No Such Thing as a Free Lunch* (LaSalle, IL: Open Court, 1975).
28. Fermor, *The Traveller's Tree*, 84.
29. Kenneth John, "The Rise of the Breadfruit," *The Vincentian*, October 4, 2019. https://thevincentian.com/the-rise-of-the-breadfruit-p18039-108.htm.
30. G. Wright, "Economic Conditions in St. Vincent, B.W.I.," *Economic Geography* 5, no. 3 (1929): 236–59, quote at 244; Hélène Bernier, Claude Sastre, and Michel Magras, *Plantes utilitaires de Saint-Barthélemy* (Gustavia, St. Barthélemy: St. Barth Essentiel, 2012), 50, translation mine.
31. My source for population data is table 4.1 at 75–76 in Pieter C. Emmer and Stanley L. Engerman, "The Non-Hispanic West Indies," in *The Cambridge World History of Slavery*, ed. David Eltis, Stanley L. Engerman, Seymour Drescher, and David Richardson (Cambridge: Cambridge University Press, 2017), 4:73–97.
32. John Thieme, *Postcolonial Literary Geographies: Out of Place* (London: Palgrave Macmillan, 2016), 51–52.
33. Gregg Mitman, "Reflections on the Plantationocene: A Conversation with Donna Haraway and Anna Tsing," *Edge Effects*, June 18, 2019, https://edgeeffects.net/wp-content/uploads/2019/06/PlantationoceneReflections_Haraway_Tsing.pdf; Hannah Rachel Cole, "Breadfruit in the Wake: Imagining Vegetal Mutiny in Derek Walcott's 'The Bounty,'" *Latin American Literary Review* 48, no. 96 (2021): 35–39.
34. Smilowitz, "Fruits of the Soil," 30; Derek Walcott, *The Bounty* (New York: Farrar, Straus and Giroux, 1997), 9.
35. Walcott, *The Bounty*, quotes throughout; Cole, "Breadfruit in the Wake," 38.
36. Smilowitz, "Fruits of the Soil," 30.

9. Fruits of the Creole Kind

1. Whitney Smith, "Flag of Saint Vincent and the Grenadines," *Encyclopaedia Britannica* (2011), https://www.britannica.com/topic/flag-of-Saint-Vincent-and-the-Grenadines. The only other national symbol to include breadfruit imagery is the Bolivian coat of arms, which has seen many stylistic changes throughout its existence. The main theme of these changes has been to replace European iconography with Indigenous symbols. For example, in 1924, then-President

Bautista Saavedra issued a decree replacing the olive and laurel with the *kantuta*, one of Bolivia's two national flowers. In many cases, though apparently still nonstandard, the (imported) breadfruit tree is replaced by a (native) palm tree. I am indebted to Julio César Velásquez Alquizaleth for his exhaustive work on the subject of historical changes to the Bolivian coat of arms as detailed in his website. According to Velásquez, the 1825 decree that established the basis for the modern coat of arms stated that "en el costado, ira grabado sobre campo blanco el árbol prodigioso denominado del pan, que se encuentra en varias de las montañas de la República, significándose por el la riqueza del Estado en el Reino, Vegetal" (on the side, to be engraved over a white field, the prodigious tree called bread, which is found in several of the mountains of the Republic, signifying the wealth of the State in the Vegetable Kingdom). Julio César Velásquez Alquizaleth, *Reformas y deformaciones del escudo de armas de la República de Bolivia*, 1998. http://www.bolivian.com/escudos/; V. S. Naipaul, *The Middle Passage: Impressions of Five Societies—British, French, and Dutch—in the West Indies and South America* (London: Readers Union/Andre Deutsch, 1963), 182; Robin Wall Kimmerer, *Braiding Sweetgrass: Indigenous Wisdom, Scientific Knowledge, and the Teachings of Plants* (Minneapolis: Milkweed, 2013), 9.

2. William Dampier, *A new Voyage round the World* (London: Knapton, 1697), 68.
3. Robert A. Hill, ed. *The Marcus Garvey and Universal Negro Improvement Association Papers* (Berkeley: University of California Press, 1990), 7:791; Alemseghed Kebede and J. David Knottnerus, "Beyond the Pales of Babylon: The Ideational Components and Social Psychological Foundations of Rastafari," *Sociological Perspectives* 41, no. 3 (1998), 499–517.
4. Kebede and Knottnerus, "Beyond the Pales," 503; Ariella Y. Werden-Greenfield, "Warriors and Prophets of Livity: Samson and Moses as Moral Exemplars in Rastafari," PhD diss. (Temple University, 2016); E. Ellis Cashmore, *Rastaman: Rastafarian Movement in England* (London: Allen and Unwin, 1980), 67.
5. Werden-Greenfield, "Warriors and Prophets." On the Nazarite vow, see Numbers 6:2–8.
6. Brien A. Meilleur, Noa Kekuewa Lincoln, Joannie Dobbs, C. Alan Titchenal, Richard R. Jones, and Alvin S. Huang, *Hawaiian Breadfruit: Ethnobotany, Human Ecology, Agronomy, Nutrition, Modernity* (Honolulu: College of Tropical Agriculture and Human Resources, University of Hawai'i-Mānoa, 2024); Gloridene Hoyte-John, "Breadfruit Wood Coffin for Sir Frederick," *iWitness News*, January 27, 2020, https://www.iwnsvg.com/2020/01/27/breadfruit-wood-coffin-for-sir-frederick/; Wendell Roelf, "Mourners Pay Respects to South Africa's Anti-Apartheid Hero Tutu," *Reuters*, December 30, 2021, https://www.reuters.com/world/africa/south-africas-anti-apartheid-hero-tutu-lie

-state-stgeorges-cathedral-2021-12-30/; Marie Catherine Neal, *In Gardens of Hawaii* (Honolulu: Bernice P. Bishop Museum, 1948).

7. Kimmerer, *Braiding Sweetgrass*, 9.
8. Antonio de Ulloa, *A Voyage to South America*, trans. John Adams (London: Davis and Reymers, 1758), 2:17.
9. "From William Bligh,", 4:111; "From Christopher Smith," 4:143; "From James Wiles," 4:180–81, all in *The Indian and Pacific Correspondence of Sir Joseph Banks, 1768–1820*, ed. Neil Chambers (London: Pickering & Chatto, 2011), edited for spelling. Wiles's 1801 letter to Banks was cited in Jennifer Newell, *Trading Nature: Tahitians, Europeans, and Ecological Exchange* (Honolulu: University of Hawai'i Press, 2010), 168.
10. "From Alexander Anderson," in Chambers, ed., *The Indian and Pacific Correspondence of Sir Joseph Banks*, 4:371–73, quote at 371.
11. Daniel Pauly, "Anecdotes and the Shifting Baseline Syndrome of Fisheries," *Trends in Ecology & Evolution* 10, no. 10 (1995): 430.
12. On the introduction of plantains and yams into the Caribbean see Cruz Miguel Ortíz Cuadra, *Eating Puerto Rico: A History of Food, Culture, and Identity* (Chapel Hill: University of North Carolina Press, 2013), and Candice Goucher, *Congotay! Congotay! A Global History of Caribbean Food* (New York: Routledge, 2015).
13. Candice Goucher, "Eating the World: The Iguana's Tale of Caribbean Ecology and Culinary History," in *Encounters Old and New in World History*, ed. Alan Karras and Laura J. Mitchell (Honolulu: University of Hawai'i Press, 2017), 91–106, quote at 97.
14. Laura B. Roberts-Nkrumah, *The Breadfruit Germplasm Collection at the University of the West Indies, St. Augustine Campus* (Mona, Jamaica: University of the West Indies Press, 2018); Oral O. Daley, Laura B. Roberts-Nkrumah, and Angela T. Alleyne, "Morphological Diversity of Breadfruit (*Artocarpus altilis* [Parkinson] Fosberg) in the Caribbean," *Scientia Horticulturae* 266 (2020): 109278.
15. On the European Union's Protected Designation of Origin laws and similar policies, see David M. Higgins, *Brands, Geographical Origin, and the Global Economy: A History from the Nineteenth Century to the Present* (Cambridge: Cambridge University Press, 2018).
16. Amanda Morris, "Caribbean Breadfruit Traced Back to Capt. Bligh's 1791–93 Journey," *Northwestern Now*, January 5, 2023, https://news.northwestern.edu/stories/2023/01/caribbean-breadfruit-traced-back-to-capt-blighs-1791-93-journey/.
17. Lauren Audi, Gordon Shallow, Erasto Robertson, Dean Bobo, Diane Ragone, Elliot M. Gardner, Babita Jhurree-Dussoruth, Jacek Wajer, and Nyree J. C. Zerega, "Linking Breadfruit Cultivar Names Across the Globe Connects Histories After 230 Years of Separation," *Current Biology* (2022): 287–97.

18. Audi et al., "Linking Breadfruit Cultivar Names," 2.
19. Lauren Audi, Nyree Zerega, Gordon Shallow, Erasto Robertson, and Diane Ragone, "Breadfruit of St. Vincent and the Grenadines," March 2022, https://laudi-ecology.com/writing-2/.
20. William J. Hooker, "*Artocarpus incisa*: Bread-Fruit Tree (α and β)," *Curtis's Botanical Magazine* 55 (1828): 2869–71, quote at 2871.

10. Roast or Fry

1. Sandra Mason, "Speech from the Throne (Bridgetown, Barbados)," Barbados Government Information Service, September 15, 2020, https://gisbarbados.gov.bb/download/throne-speech-delivered-september-15-2020/.
2. The Errol Barrow quote that Mason referenced is found in Shridath Ramphal's 2009 Errol Barrow Memorial Lecture, titled "Glimpses of the Commonwealth at Sixty," the text of which can be read at http://www.caribbeanelections.com/eDocs/articles/bb/Errol%20Barrow%20Memorial%20Lecture%202009.pdf.
3. Mason, "Speech from the Throne."
4. Patrick Leigh Fermor, *The Traveller's Tree: A Journey Through the Caribbean Islands* (London: John Murray, 1950), 136.
5. Austin Clarke, *Pig Tails 'n Breadfruit: Rituals of Slave Food, a Barbadian Memoir* (Toronto: Vintage Canada, 2000), 115.
6. On the decline of the sugar industry in Barbados, see Ian Drummond and Terry Marsden, "A Case Study of Unsustainability: The Barbados Sugar Industry," *Geography* 80, no. 4 (1995): 342–54.
7. The quotes from Carmeta Fraser come from an undated advertisement for the BADMC titled, "If we can produce it here, don't import it." The advertisement is no longer available through the website of the BADMC but can be found online via the Internet Archive at https://web.archive.org/web/20220122142311/https://badmc.org/wp-content/uploads/2016/03/jpg2pdf.pdf.
8. "Three men die in separate breadfruit tree incidents within same week," *Loop News*, March 25, 2025, https://www.loopnews.com/content/three-men-die-in-separate-breadfruit-tree-incidents-within-same-week/.
9. Many published versions of the stone soup story exist, and its origin is lost to history. See, for example, Marcia Brown, *Stone Soup: An Old Tale* (New York: Scribner, 1947).
10. On the precarious food and economic situation of St. Barthélemy, see Michael Gross, "Clouds on the Horizon in St. Barth," *Palmer*, August 8, 2023, https://palmerpb.com/2023/08/08/st-barth/.

11. On the sofi (specialty outstanding food innovation) Award, see "sofi™ Award Winners," Specialty Food Association, https://www.specialtyfood.com/awards/sofi-awards/. "KeHE® Selects 44 'Golden Ticket' Winners at Latest TRENDfinder™ Event," *KeHE*, November 30, 2023, https://www.kehe.com/news-blog/news/kehe-distributors-trend-finder%E2%88%92event-golden-ticket-winners/.
12. Puerto Rican Spanish includes at least three terms for breadfruit: *pana*, *panapén*, and *mapén*.
13. On the deaths attributed to Hurricane Maria, see Carlos Santos-Burgoa, Ann Goldman, Elizabeth Andrade, Nicole Barrett, Uriyoan Colon-Ramos, Mark Edberg, Alejandra Garcia-Meza, Lynn Goldman, Amira Roess, John Sandberg, and Scott Zeger, *Ascertainment of the Estimated Excess Mortality from Hurricane María in Puerto Rico* (Washington, DC: George Washington University, 2018), available online at https://publichealth.gwu.edu/sites/g/files/zaxdzs4586/files/2023-06/acertainment-of-the-estimated-excess-mortality-from-hurricane-maria-in-puerto-rico.pdf. On the monetary damages attributed to the storm, see Eric S. Blake, "The 2017 Hurricane Season: Catastrophic Losses and Costs," *Weatherwise* 71, no. 3 (2018): 28–37.
14. "New Flags: Netherlands Antilles, St. Vincent and the Grenadines," *Flag Bulletin* 25, no. 6 (1986): 203–8.
15. Roman Mars, "Why City Flags May Be the Worst-Designed Thing You Never Noticed," TED, May 2015, https://www.ted.com/talks/roman_mars_why_city_flags_may_be_the_worst_designed_thing_you_ve_never_noticed.
16. Chi Ching Ching, "Roast or Fry (Breadfruit)," on *Turning Tables* (Dutty Rock Productions, 2018).
17. Chi Ching Ching, "Roast or Fry (Breadfruit)," YouTube, March 4, 2016, https://www.youtube.com/watch?v=IGdtxQM8TuI.

11. The Second-Best Time to Plant a Tree

1. Richard F. Burton, *Wanderings in West Africa: From Liverpool to Fernando Po* (London: Tinsley Brothers, 1863), 2:223; J. M. Dalziel, *The Useful Plants of West Tropical Africa* (London: The Crown Agents for the Colonies, 1948), 274; Stanley B. Alpern, "The European Introduction of Crops into West Africa in Precolonial Times," *History in Africa* 19 (1992): 13–43.
2. Carl Christian Reindorf, *History of the Gold Coast and Asante* (Basel: Missionbuchhandlung, 1895), 269.
3. Matthew 25:14–30 (KJV).

4. United Nations, *Transforming Our World: The 2030 Agenda for Sustainable Development*, A/RES/70/1. https://sdgs.un.org/publications/transforming-our-world-2030-agenda-sustainable-development-17981.
5. George T. Bagby, "Report on the Effects of Channelization and Draining by the US Soil Conservation Service in Watersheds in Georgia," *Congressional Record* 115, part 29 (1969): 38980–81, quote at 38981.
6. Stefano Padulosi, Judith Thompson, and Per Rudebjer, *Fighting Poverty, Hunger and Malnutrition with Neglected and Underutilized Species (NUS): Needs, Challenges and the Way Forward* (Rome: Bioversity International, 2013); Diane Ragone, *Breadfruit (*Artocarpus altilis *[Parkinson] Fosberg): Promoting the Conservation and Use of Underutilized and Neglected Crops* (Rome: International Plant Genetic Resources Institute, 1997); Food and Agricultural Organization of the United Nations, "Annex I: List of Crops Covered Under the Multilateral System," *The Multilateral System*, https://www.fao.org/plant-treaty/areas-of-work/the-multilateral-system/annex1/en/.
7. Karolina M. Zarzyczny, Marc Rius, Suzanne T. Williams, and Phillip B. Fenberg, "The Ecological and Evolutionary Consequences of Tropicalisation," *Trends in Ecology & Evolution* 39, no. 3 (2024): 267–79; M. F. Cardell, A. Amengual, and R. Romero, "Future Effects of Climate Change on the Suitability of Wine Grape Production Across Europe," *Regional Environmental Change* 19 (2019): 2299–310, quote at 2299.
8. Lucy Yang, Nyree Zerega, Anastasia Montgomery, and Daniel E. Horton, "Potential of Breadfruit Cultivation to Contribute to Climate-Resilient Low Latitude Food Systems," *PLOS Climate* 1, no. 8 (2022): e0000062; Sarah Kutah, "Is Breadfruit the Climate Change-Proof Food of the Future?," *Smithsonian*, August 30, 2022, https://www.smithsonianmag.com/smart-news/is-breadfruit-the-climate-change-proof-food-of-the-future-180980665/.
9. Chad Livingston and Noa Kekuewa Lincoln, "Determining Allometry and Carbon Sequestration Potential of Breadfruit (*Artocarpus altilis*) as a Climate-Smart Staple in Hawai'i," *Sustainability* 15 (2023): 15682.
10. Both potatoes and rice are technically perennial crops but are almost always plowed up and replanted each year—that is, treated as annuals—owing to the reduced yield in secondary seasons and beyond. Agricultural scientists are working to develop high-yield perennial rice and some farmers report success with growing potatoes as a perennial.
11. Ross Barrett, "Picturing a Crude Past: Primitivism, Public Art, and Corporate Oil Promotion in the United States," *Journal of American Studies* 46, no. 2 (2012): 395–422; see also Sinclair Oil, "DINO History," https://www.sinclairoil.com/dino-history.
12. Jeffrey S. Dukes, "Burning Buried Sunshine: Human Consumption of Ancient Solar Energy," *Climatic Change* 61 (2003): 31–44.

13. Christopher McDonough, "Oil and Water," *Immanence* 3, no. 2 (2019): 11–15, quotes throughout, edited for language.
14. McDonough, "Oil and Water," quotes throughout, edited for language.
15. David Biello, "The Origin of Oxygen in Earth's Atmosphere," *Scientific American*, August 19, 2019, https://www.scientificamerican.com/article/origin-of-oxygen-in-atmosphere/; Ella Frances Sanders, *Eating the Sun: Small Musings on a Vast Universe* (New York: Penguin, 2019), 3.
16. Robert Costanza, Ralph d'Arge, Rudolf de Groot, Stephen Farber, Monica Grasso, Bruce Hannon, Karin Limburg, Shahid Naeem, Robert V. O'Neill, Jose Paruelo, Robert G. Raskin, Paul Sutton, and Marjan van den Belt, "The Value of the World's Ecosystem Services and Natural Capital," *Nature* 387 (1997): 253–60; "Ecosystem Services & Biodiversity (ESB)," Food and Agricultural Organization of the United Nations, http://www.fao.org/ecosystem-services-biodiversity/en/.
17. Robert Costanza, Rudolf de Groot, Paul Sutton, Sander van der Ploeg, Sharolyn J. Anderson, Ida Kubiszewseki, Stephen Farber, and R. Kerry Turner, "Changes in the Global Value of Ecosystem Services," *Global Environmental Change* 26 (2014): 152–58.
18. Evidence for the earliest known use of numbers comes from the Lebombo bone, a baboon fibula used as a tally stick and determined to be about 37,000 years old. The bone features twenty-nine marks, prompting researchers to conclude that its use involved tracking lunar cycles. An interesting alternative (or additional) interpretation, based on the assumption that "women would benefit from keeping track of the menstrual cycle, which requires a lunar calendar," concludes that "the first mathematicians probably were women." Since it would take about 31,689 years to count to one trillion at one integer per second, humans have barely—relatively speaking—used numbers long enough to have counted to a trillion. On number history, see Johanna Pejlare and Kajsa Bråting, "Writing the History of Mathematics: Interpretations of the Mathematics of the Past and Its Relation to the Mathematics of Today," in *Handbook of the Mathematics of the Arts and Sciences*, ed. Bharath Sriraman (Cham, Switzerland: Springer, 2021), 2395–420, quotes cited in this note at 2398. On the Lebombo bone, see Francesco d'Errico, Lucinda Backwell, Paola Villa, Ilaria Degano, Jeannette J. Lucejko, Marion K. Bamford, Thomas F. G. Higham, Maria Perla Colombini, and Peter B. Beaumont, "Early Evidence of San Material Culture Represented by Organic Artifacts from Border Cave, South Africa," *Proceedings of the National Academy of Sciences* 109, no. 33 (2012): 13214–19. On the use of tally sticks in general, see Sture Lagercrantz, "Counting by Means of Tally Sticks or Cuts on the Body in Africa," *Anthropos* 68 (1973): 569–88. Finally, for evidence of what may be a counting system even older than the Lebombo bone, from the famous cave paintings of southern

Europe, see Bennett Bacon, Azadeh Khatiri, James Palmer, Tony Freeth, Paul Pettitt and Robert Kentridge, "An Upper Palaeolithic Proto-Writing System and Phenological Calendar," *Cambridge Archaeological Journal* 33, no. 3 (2023): 371–89.

19. Adam Smith, *An Inquiry Into the Nature and Causes of the Wealth of Nations* (London: Strahan and Cadell, 1776), 1:441. On the history of natural capital, see the blog of Policy Ambiental, "Cashing in on Conservation: A History of Payments for Ecosystem Services," September 1, 2017, https://www.policy-ambiental.org/blog/2017/8/31/payment-for-ecosystems-services.
20. Bill Gates, "Climate Change and the 75% Problem," *GatesNotes*, October 17, 2018, https://www.gatesnotes.com/My-plan-for-fighting-climate-change.
21. On the role of New York City's tap water in bagel and pizza dough, see Matt Blitz, "Is New York Water Really the Secret to the Best Bagels and Pizza?," *Food & Wine*, March 14, 2023, https://www.foodandwine.com/news/new-york-water-bagels-pizza.
22. On the use of payments for ecosystem services in the New York watershed conservation program, see David Soll, *Empire of Water: An Environmental and Political History of the New York City Water Supply* (Ithaca, NY: Cornell University Press, 2013); National Academies of Sciences, Engineering, and Medicine, *Review of the New York City Watershed Protection Program* (Washington, DC: National Academies Press, 2020); and Fanyuan Lin, "A Comparative Study on Payment Schemes for Watershed Services in New York City and Beijing," *Consilience: The Journal of Sustainable Development* 11, no. 1 (2014): 27–40.

12. Super Food

1. Matt Fitzgerald, *Diet Cults: The Surprising Fallacy at the Core of Nutrition Fads and a Guide to Healthy Eating for the Rest of Us* (New York: Simon & Schuster, 2014), 104. Published lists of alleged superfoods are ubiquitous. Those included here were found in the following sources: Katherine D. McManus, "10 Superfoods to Boost a Healthy Diet," *Harvard Health Blog*, August 29, 2018, https://www.health.harvard.edu/blog/10-superfoods-to-boost-a-healthy-diet-2018082914463; Amy Sowder and Jen Wheeler, "What Are Superfoods, Exactly?," *CNET Health and Wellness*, August 22, 2019, https://www.cnet.com/health/what-are-superfoods-exactly/; Kamal Niaz, Elizabeta Zaplatic, and Jonathan Spoor, "Highlight Report: *Diploptera functata* (Cockroach) Milk as Next Superfood," *EXCLI* 17 (2018): 721–23.
2. Melinda Butterworth, Georgia Davis, Kristina Bishop, Luz Reyna, and Alyssa Rhodes, "What Is a Superfood Anyway? Six Key Ingredients for Making a

Food 'Super,'" *Gastronomica: The Journal for Food Studies* 20, no. 1 (2020): 46–58.

3. Michael Pollan, *The Omnivore's Dilemma: A Natural History of Four Meals* (New York: Penguin, 2006).

4. Joseph A. Knight, "Free Radicals: Their History and Current Status in Aging and Disease," *Annals of Clinical and Laboratory Science* 28, no. 6 (1998): 331–46; Bruce N. Ames, Mark K. Shgenaga, and Tory M. Hagen, "Oxidants, Antioxidants, and the Degenerative Diseases of Aging," *Proceedings of the National Academy of Sciences of the USA* 90, no. 17 (1993): 7915–22.

5. Li Fu, Bo-Tao Xu, Xiang-Rong Xu, Ren-You Gan, Yuan Zhang, En-Qin Xia, and Hua-Bin Li, "Antioxidant Capacities and Total Phenolic Contents of 62 Fruits," *Food Chemistry* 129, no. 2 (2011): 345–50; Tao Wu, Jun Yan, Ronghua Liu, Massimo F. Marcone, Haji Akber Aisa, and Rong Tsao, "Optimization of Microwave-Assisted Extraction of Phenolics from Potato and Its Downstream Waste Using Orthogonal Array Design," *Food Chemistry* 133, no. 4 (2012): 1292–98; Yayuan Tang, Weixi Cai, and Baojun Xu, "Profiles of Phenolics, Carotenoids and Antioxidative Capacities of Thermal Processed White, Yellow, Orange and Purple Sweet Potatoes Grown in Guilin, China," *Food Science and Human Wellness* 4, no. 3 (2015): 123–32; Toilibou Soifoini, Dario Donno, Victor Jeannoda, Ernest Rakotoniaina, Soule Hamidou, Said Mohamed Achmet, Noe Rene Solo, Kamaleddine Afraitane, Cristina Giacoma, and Gabriele Loris Beccaro, "Bioactive Compounds, Nutritional Traits, and Antioxidant Properties of *Artocarpus altilis* (Parkinson) Fruits: Exploiting a Potential Functional Food for Food Security on the Comoros Islands," *Journal of Food Quality* 2018 (2018): 5697928.

6. Horace D. Graham and Negron de Bravo, "Composition of the Breadfruit," *Journal of Food Science* 46 (1981): 535–39; Neela Badrie and Jacklyn Broomes, "Beneficial Uses of Breadfruit (Artocarpus altilis): Nutritional, Medicinal and Others," in *Bioactive Foods in Promoting Health: Fruits and Vegetables*, ed. Ronald Watson and Victor R. Preedy (London: Elsevier, 2010), 491–506; Ying Liu, Paula N. Brown, Diane Ragone, Deanna L. Gibson, and Susan J. Murch, "Breadfruit Flour Is a Healthy Option for Modern Foods and Food Security," *PLoS One* 15, no. 7 (2020): e0236300.

7. Natalie Pitchford Levy and Hinnerk von Bargen, "Protein-Based Foods: Vegetable Sources of Protein and Protein Complementation," in *Culinology: The Intersection of Culinary Art and Food Science*, ed. Jeffrey Cousminer (Hoboken, NJ: Wiley, 2016), 106–27.

8. Benjamin Lebwohl, David S. Sanders, and Peter H. R. Green, "Coeliac Disease," *Lancet* 399 (2018): 70–81; Simona Gatti, Alberto Rubio-Tapia, Govind Makharia, and Carlo Catassi, "Patient and Community Health Global Burden in a World with More Celiac Disease," *Gastroenterology* 167, no. 1 (2024):

25–33; Mayo Clinic, "Gluten-Free Diet," *Nutrition and Healthy Eating*, December 11, 2021, https://www.mayoclinic.org/healthy-lifestyle/nutrition-and-healthy-eating/in-depth/gluten-free-diet/art-20048530.

9. Christine Byrne, "A Curious Eater's Guide to Alternative Flours," *Outside*, December 15, 2020, https://www.outsideonline.com/2419342/alternative-flours-gluten-free.

10. Kelsey Nowakowski, "Can Breadfruit Overcome Its Past to Be a Superfood of the Future?," *National Geographic* July 22, 2015, https://www.nationalgeographic.com/culture/food/the-plate/2015/07/22/can-breadfruit-overcome-its-past-to-be-a-superfood-of-the-future/; Rebecca Rupp, "Breadfruit and 'The Bounty' That Brought It Across the Ocean," *National Geographic*, April 28, 2016, https://www.nationalgeographic.com/culture/food/the-plate/2016/04/28/breadfruit-and-the-bounty-that-brought-it-across-the-ocean/; Laura Kiniry, "The Island Fruit That Caused a Mutiny," *BBC Travel*, May 18, 2018, http://www.bbc.com/travel/story/20180517-the-island-fruit-that-caused-a-mutiny; Josh Schonwald, "Forget Kale: Try These Three REAL Superfoods," *Time*, October 28, 2014, https://time.com/3544425/superfoods-moringa-tree-breadfruit-prickly-pear-cactus/; Derrick B. Jelliffe, "Parallel Food Classifications in Developing and Industrialized Countries," *American Journal of Clinical Nutrition* 20, no. 3 (1967): 279–81, quote at 279.

11. Sven-Erik Jacobsen, "The Worldwide Potential for Quinoa (*Chenopodium quinoa* Willd.)," *Food Reviews International* 19, no. 1–2 (2003): 167–77, quote at 167; Fabiana Li, "Materiality and the Politics of Seeds in the Global Expansion of Quinoa," *Food Culture & Society* 26, no. 4 (2023): 867–85, quote at 867; Marygold Walsh-Dilley, "Tensions of Resilience: Collective Property, Individual Gain and the Emergent Conflicts of the Quinoa Boom," *Resilience* (2015): 1094168.

12. Julia Flynn Siler, "'Food of the Future' Has One Hitch: It's All But Inedible," *Wall Street Journal*, November 1, 2011, https://www.wsj.com/articles/SB10001424052970203752604576645242121126386; Wallace quoted in Marston Bates, *Where Winter Never Comes: A Study of Man and Nature in the Tropics* (New York: Scribner, 1952), 164; Gloria Dickie, "The Last Tree Standing," *Modern Farmer*, November 5, 2018, https://modernfarmer.com/2018/11/the-last-tree-standing/.

13. Amanda Fiegl, "Breadfruit, the Holy Grail of Grocery Shopping," *Smithsonian*, August 25, 2009, https://www.smithsonianmag.com/arts-culture/breadfruit-the-holy-grail-of-grocery-shopping-66539958/; Rachel Laudan, "Breadfruit. All But Inedible?," *A Historian's Take on Food and Politics*, November 2, 2011, https://www.rachellaudan.com/2011/11/breadfruit-all-but-inedible.html; Wallace quoted in Bates, *Where Winter Never Comes*, 164.

14. Fiegl, "Breadfruit"; James McWilliams, "Fruit Mutiny," *Paris Review Daily*, August 1, 2014, https://www.theparisreview.org/blog/2014/08/01/fruit-mutiny/.

15. Jean-Jacques Rousseau, *The Social Contract & Discourses*, trans. G. D. H. Cole (London: Dent, 1920), 207.
16. Daphne Ewing-Chow, "Breadfruit Has All the Makings of a Global Future Food Trend," *Forbes*, September 30, 2023, https://www.forbes.com/sites/daphneewingchow/2023/09/30/breadfruit-is-a-climate-smart-superfood-with-global-appeal/; Kim Severson, "9 Predictions for How We'll Eat in 2024," *New York Times*, December 26, 2023, https://www.nytimes.com/2023/12/26/dining/food-trends-2024.html; AF & Co. and Carbonate, "Hospitality Trends Report," *Carbonate*, https://www.carbonategroup.com/trend-report.
17. Otis Warren Barrett, *The Tropical Crops: A Popular Treatment of the Practice of Agriculture in Tropical Regions, with Discussion of Cropping Systems and Methods of Growing the Leading Products* (New York: Macmillan, 1928), 198.
18. Amy Stewart, *The Drunken Botanist: The Plants That Create the World's Great Drinks* (Chapel Hill, NC: Algonquin Books, 2013), xiv.
19. U.S. Department of the Treasury, Alcohol & Tobacco Tax & Trade Bureau, *The Beverage Alcohol Manual (BAM): A Practical Guide*, vol. 2, *Basic Mandatory Labeling Information for Distilled Spirits* (Washington, DC: Department of the Treasury, 2007), 4–2.

13. Pushing Latitude

1. Charles E. Fairman, *Art and Artists of the Capitol of the United States of America* (Washington, DC: Government Printing Office, 1927), 222.
2. April Rubin, "Statue of Black Educator Replaces Confederate General in U.S. Capitol," *New York Times*, July 13, 2022; Randy Lee Loftis, "Forty More Years of Crisis," in Marjory Stoneman Douglas, *The Everglades: River of Grass* (Sarasota, FL: Pineapple Press, 1997), 404; Minna Scherlinder Morse, "Chilly Reception," *Smithsonian*, July 2002, https://www.smithsonianmag.com/history/chilly-reception-66099329/. On Gorrie, see Linda Caldwell, *He Made Ice and Changed the World: The Story of Florida's John Gorrie* (Ocala, FL: Atlantic, 2019) and Raymond Arsenault, "The Cooling of the South," *Wilson Quarterly* 8, no. 3 (1984): 150–59.
3. David Dondero, "South of the South," on *South of the South* (Team Love Records, 2005).
4. Agricultural Research Service, U.S. Department of Agriculture, USDA Plant Hardiness Zone Map, 2023, https://planthardiness.ars.usda.gov/.
5. Julia F. Morton, *Fruits of Warm Climates* (Miami: Julia F. Morton, 1987), 50–58, available online at https://www.hort.purdue.edu/newcrop/morton/index.html; Patrick Breen, *"Artocarpus altilis,"* Oregon State University: Landscape

Plants, 2024, https://landscapeplants.oregonstate.edu/node/2146; Jonathan Crane, associate director of the Tropical Research & Education Center at the University of Florida, email to author, February 23, 2021; Pliny Reasoner and Egbert Reasoner, *Descriptive and Illustrated Catalogue and Manual of Royal Palm Nurseries, Oneco, Florida, U.S.A.* (Harrisburg, PA: McFarland, 1892), 7.

6. Herbert S. Wolfe, "Fifty Years of Tropical Fruit Culture," *Proceedings of the Florida State Horticultural Society* 50 (1937): 72–78, quote at 76.

7. Colin L. A. Leakey, *Breadfruit Reconnaissance Study in the Caribbean Region* (Cali, Colombia: CIAT, 1977); Colin Leakey, "Biography," *Colin Leakey*, https://sites.google.com/site/colinleakey/biography/.

8. National Oceanic and Atmospheric Administration (NOAA), "State of the Climate: Global Climate Report for Annual 2020," National Centers for Environmental Information, January 2021, https://www.ncdc.noaa.gov/sotc/global/202013; Anthony Arguez, Shannan Hurley, Anand Inamdar, Laurel Mahoney, Ahira Sanchez-Lugo, and Lilian Yang, "Should We Expect Each Year in the Next Decade (2019–28) to Be Ranked Among the Top 10 Warmest Years Globally?," *Bulletin of the American Meteorological Society* 101, no. 5 (2020), e655–e663.

9. "From Benjamin Vaughn," in *The Papers of Thomas Jefferson*, ed. Julian P. Boyd (Princeton, NJ: Princeton University Press, 1958), 14:673–74. On the presence of breadnut and the absence of seeded cultivars of breadfruit in the Caribbean, see Guylène Aurore, Joselle Nacitas, Berthe Parfait, and Louis Fahrasmane, "Seeded Breadfruit Naturalized in the Caribbean Is Not a Seeded Variety of *Artocarpus altilis*," *Genetic Resources and Crop Evolution* 61 (2014): 901–7.

10. "From Benjamin Vaughan," with enclosure: "An Account of the Mutiny on the *Bounty*," 16:274–76, and "To Benjamin Vaughan," 15:133–34, in Boyd, ed., *The Papers of Thomas Jefferson*; "From Fairlie Christie," in *The Papers of George Washington*, ed. David R. Hoth and Carol S. Ebel (Charlottesville: University of Virginia Press, 2013), 17:682–85; "To Fairlie Christie, 25 May 1795," in *The Papers of George Washington*, ed. Carol S. Ebel (Charlottesville: University of Virginia Press, 2015), 18:171; George Washington Papers, Series 4, *General Correspondence: George Washington to Thomas Newton Jr.* 1795. Manuscript/Mixed Material, https://www.loc.gov/item/mgw438675/.

11. Erica R. Johnson, *Philanthropy and Race in the Haitian Revolution* (New York: Palgrave Macmillan, 2018); "From Alexandre Giroud," in *The Papers of Thomas Jefferson*, ed. Barbara B. Oberg (Princeton, NJ: Princeton University Press, 2002), 29:347–48. On the political undertones of Giroud's gift to Jefferson, see Christopher P. Iannini, *Fatal Revolutions: Natural History, West Indian Slavery, and the Routes of American Literature* (Chapel Hill: University of North Carolina Press, 2012), ch. 5.

12. "To Alexandre Giroud," in Oberg, ed., *The Papers of Thomas Jefferson*, 29:387–88.

13. "To Allen Jones," in Oberg, ed., *The Papers of Thomas Jefferson*, 29:388; Philippa Mein Smith, *A Concise History of New Zealand* (Cambridge: Cambridge University Press, 2012), 18.
14. "From Allen Jones," 29:513–14, and "From Thomas Bee," 29:487, in Oberg, ed., *The Papers of Thomas Jefferson*.
15. "From Samuel Maverick," 18:257, and "To Samuel Maverick," 18:378, in *The Papers of Thomas Jefferson*, ed. J. Jefferson Looney (Princeton, NJ: Princeton University Press, 2021). On Samuel Maverick, see National Public Radio, "Original 'Maverick' Was Unconventional Texan," *Morning Edition*, September 5, 2008, https://www.npr.org/templates/story/story.php?storyId=94312345.
16. William Francis Whitman, *Five Decades with Tropical Fruit: a Personal Journey*, ed. Donna McVicar Cannon (Englewood, FL: Quisqualis, 2001), quotes throughout.
17. Whitman, *Five Decades*, quotes throughout.
18. Technically, "NS-1" was not the first jackfruit cultivar introduced into the United States. According to the U.S. Department of Agriculture, "NS-1" arrived at Miami in June 1978, and horticulturist Julia Morton has written that the species was in Florida by the early 1880s. "NS-1" seems, however, to have been the first named cultivar officially brought into the United States with documentation from the USDA. Morton, *Fruits of Warm Climates*, 58–64; USDA National Plant Germplasm System, "Details for: MIA 25128, *Artocarpus heterophyllus* Lam., NA N.S. 1," *Agriculture Research Service*, https://npgsweb.ars-grin.gov/gringlobal/accessiondetail?id=1090976.
19. Whitman, *Five Decades*, 256; NOAA and National Weather Service (NWS), "Summary of Historic Cold Episode of January 2010: Coldest 12-Day Period Since at Least 1940," weather.gov, https://www.weather.gov/media/mfl/news/ColdEpisodeJan2010.pdf; NOAA and NWS, "Climatological Records for Miami, FL," weather.gov, https://www.weather.gov/media/mfl/climate/Daily_Records_Miami.pdf.
20. Kalisi Mausio, Tomoaki Miura, and Noa K. Lincoln, "Cultivation Potential Projections of Breadfruit (*Artocarpus altilis*) Under Climate Change Scenarios Using an Empirically Validated Suitability Model Calibrated in Hawai'i," *PLoS ONE* 15, no. 5 (2020): e0228552.
21. "To Alexandre Giroud."
22. Patrick B. Garvey, "Saving the Grimal Grove, Restoring a Legendary Tropical Fruit Collection and Establishing Grimal Grove as the Southernmost Tropical Fruit Park," *Proceedings of the Florida Horticultural Society* 129 (2016): 35–36; Whitman, *Five Decades*, 347.
23. Tiffany Duong. "Growing Hope—Historic Grimal Grove Re-opening, Becoming 'Grove of the Future,'" *Keys Weekly*, November 8, 2019, https://keys

weekly.com/42/growing-hope-historic-grimal-grove-re-opening-becoming-grove-of-the-future/.

24. Like Washington, DC, the Australian capital, Canberra, is not located within any state; unlike DC, Canberra is placed within its own designated territory: the Australian Capital Territory, or ACT. So that this landlocked territory might have access to the sea, another territory—Jervis Bay—has been established to be administered as if it were a part of the ACT. Both the ACT and Jervis Bay Territory are much smaller than any other Australian state or territory, at 1,465 square miles (2,358 km²) and 41.6 square miles (67 km²), respectively; the next smallest state is Tasmania at 40,090 square miles (64,519 km²). Data on the areas of Australian states and territories were gathered from "Area of Australia—States and Territories," *Geoscience Australia*, July 26, 2023. https://www.ga.gov.au/scientific-topics/national-location-information/dimensions/area-of-australia-states-and-territories.

25. On cassowaries and the dangers they can present to humans, see Christopher P. Kofron, "Attacks to Humans and Domestic Animals by the Southern Cassowary (*Casuarius casuarius johnsonii*) in Queensland, Australia," *Journal of Zoology* 249, no. 4 (1999): 375–81; Darren Naish, "How Dangerous Are Cassowaries, Really?," *Scientific American*, June 8, 2016. https://blogs.scientificamerican.com/tetrapod-zoology/how-dangerous-are-cassowaries-really/.

26. Roger Goebel, "Breadfruit—the Australian Scene," *Acta Horticulturae* 757 (2007): 141–48.

27. Phil Crumbley, "Betsy and the Breadfruit Tree" (unpublished, undated manuscript), PDF.

28. Tom Whitney, email to author, March 5, 2024.

29. Russell Fielding and Jorge Zaldivar, "No Longer 'Confined to the Lower Keys of Florida': Mainland United States Cultivation of Breadfruit (*Artocarpus altilis*) in a Changing Climate," *Annals of the American Association of Geographers* 113, no. 2 (2022): 370–89; Kimberly Miller, "Breadfruit Is Growing in PBC—but Shouldn't Be," *Palm Beach Post*, August 13, 2023.

14. Two Trees

1. Revelation 22:2 (KJV).
2. Mark Makowiecki, "Untangled Branches: The Edenic Tree(s) and the Multivocal *WAW*," *Journal of Theological Studies*, NS 71, no. 2 (2020), 441–57, quotes at 441 and 447, emphasis in the original; Genesis 3:7 (KJV).
3. Makowiecki, "Untangled Branches," 442, 450–51.

4. D. Zaiger, "The Tree of Life Is Dying," *Micronesian Reporter* 15, no. 1 (1967): 30–34. On the applications of Western theological concepts to breadfruit in Pacific contexts, see Vanessa Smith, "Give Us Our Daily Breadfruit: Bread Substitution in the Pacific in the Eighteenth Century," *Studies in Eighteenth Century Culture* 35 (2006): 53–75; Craig Elevitch, Diane Ragone, and Ian Cole, *Breadfruit Production Guide: Recommended Practices for Growing, Harvesting, and Handling* (Kalaheo, HI: Breadfruit Institute of the National Tropical Botanical Garden, 2014).

5. Jamaica Kincaid, *My Garden (Book)* (New York: Farrar Straus Giroux, 1999), 136–37; Karim Ganem Maloof, "The Breadfruit Tree," *Strangers Guide* 10 (2021): 62–65, quote at 62. On Colombian slavery, see William Frederick Sharp, *Slavery on the Spanish Frontier: The Colombian Chocó, 1680–1810* (Norman: University of Oklahoma Press, 1976), and on the process of emancipation there, see Edgardo Pérez Morales, *Unraveling Abolition: Legal Culture and Slave Emancipation in Colombia* (Cambridge: Cambridge University Press, 2022).

6. "About the Project," *The Pōpolo Project*, https://www.thepopoloproject.org/home. Eric Stinton, "A Word, a Plant, a Group of People: Unpacking 'Pōpolo,'" *KHON2*, February 1, 2020, https://www.khon2.com/hidden-history/black-history-month/a-word-a-plant-a-group-of-people-unpacking-popolo.

7. Lauren Audi, Gordon Shallow, Erasto Robertson, Dean Bobo, Diane Ragone, Elliot M. Gardner, Babita Jhurree-Dussoruth, Jacek Wajer, and Nyree J. C. Zerega, "Linking Breadfruit Cultivar Names Across the Globe Connects Histories After 230 Years of Separation," *Current Biology* (2022): 287–97.

8. Kincaid, *My Garden (Book)*, 136; James Mitchell, *Beyond the Islands: An Autobiography* (London: Macmillan Caribbean, 2006), 444.

9. "Let Us Beat Swords Into Ploughshares," *United Nations Gifts*, https://www.un.org/ungifts/let-us-beat-swords-ploughshares; Isaiah 2:4 (KJV), edited for spelling.

10. Elizabeth M. DeLoughrey, "Globalizing the Routes of Breadfruit and Other Bounties," *Journal of Colonialism and Colonial History* 8, no. 3 (2007), para. 45; Hannah Rachel Cole, "Breadfruit in the Wake: Imagining Vegetal Mutiny in Derek Walcott's 'The Bounty,'" *Latin American Literary Review* 48, no. 96 (2021): 35–39.

11. At the time of writing, CARICOM's Ten-Point Plan is not available on the organization's own website. It can be accessed, however, via the United Nations Human Rights Office at https://adsdatabase.ohchr.org/IssueLibrary/CARICOM_Ten-Point%20Plan%20for%20Reparatory%20Justice.pdf.

12. CARICOM, Ten Point Plan; Charles Griswold, *Forgiveness: A Philosophical Exploration* (Cambridge: Cambridge University Press, 2007), 149.

13. CARICOM, Ten Point Plan; Roy L. Brooks, *Atonement and Forgiveness: A New Model for Black Reparations* (Berkeley: University of California Press, 2004), x; Myisha Cherry, "Racialized Forgiveness," *Hypatia* 36 (2021): 583–97.
14. Ludwig Wittgenstein, *Philosophical Investigations*, trans. G. E. M. Anscombe (New York: Macmillan, 1958), 223.
15. Jonathan Drori, *Around the World in 80 Trees* (London: Laurence King, 2018).
16. Makowiecki, "Untangled Branches," 441.

Index

Page numbers in italics refer to figures.

Abankwa, Victoria, 170, 180
Abbey, Samuel Glory, 164–66, 180
Acharya Jagadish Chandra Bose Indian Botanic Garden, 96
Adams, John (*Bounty* sailor), 104
adventitious roots, 34–35
African breadfruit (*Treculia africana*), 210–11
Agwani, Kwesi, 168, 170, 179, 194–95
Ah-Scha, Hervé, 13–14, *16*, 65
Aimard, Gustave, 106
air-layering, 34–35, 53, 97, 140–41, 178, 205, 209, 215, 217, 219–20
Alatas, Syed Hussein, 62
Alfred P. Sloan Foundation, xiii–xiv
Alleyne, Chris, 149–50
Álvares, Jorge, 55
Amaral, Carlos, 217
Anderson, Alexander, 116, 135
Andía y Varela, José de, 72
Anson, George, 58–59, 66, 113

antioxidants, 183
Aotearoa (New Zealand), *2*, 45–47, 61–63, 76, 94, 202
Arouet, François-Marie (Voltaire). *See* Voltaire
Artocarpus altilis, xvi, 23–24, 231. *See also* breadfruit
Artocarpus heterophyllus, 23. *See also* jackfruit
Audi, Lauren, 141–42
Australia, xiv, *2*, 167, 211–14
Austronesians, 43–44
Avegalio, Tusi, 5

Bahamas, xiv, *90*, 156–57
Balboa, Vasco Núñez de, 55
Ballantyne, Frederick, 134
Banks, Joseph, 5–6, 96, 102, 109–12, 115, 135, 217
Barbados, *25*, *91*, 93, 107, 125, 134, 139, 143–51, *149*, 167

[277]

Barbados Agricultural Development and Marketing Corporation (BADMC), 145–47
Barrie, J. M., 64
Barrow, Errol, 93, 143–244
Beckford, Radion Tashaman (a.k.a. Chi Ching Ching). *See* Chi Ching Ching
Beckles, Hilary, 119
Bee, Thomas, 202
Bermuda, xiv, 25, 215, 217, *218*, 223
Berry, Wendell, 85
Bevacqua, Bob, 86–87
Beyuo, Christopher, 179
Bezares, Liannette, 154–55
Bismarck Islands, 39, 43–46
Blackman-Francis, Sonia, 145–47
Blake, Javion, 155–56
Bligh, William, 98–102, *101*, *103*, 127, 135, 140–42
Borneo, xiv, 40–41, *42*
Boswell, James, 124
Bougainville, Louis-Antoine de, 59–60, 64, 72
Bounty. See HMS *Bounty*
Brando, Marlon, 67–70
Brando, Teihotu, 69–70
Braun, Juliane, 113
breadfruit ('ulu): adventitious roots, 34–35; agricultural development, 106–8; *Artocarpus altilis*, xvi, 23–24, 231; as bread, 123–25; British acquisition of, xii, 75, 92–93, 96–112, 119–20, 124, 129; celebrity chefs and, 79–80; climate zones and, 196–218; colors of, xi, 14, 18–19, 24, 26–27, 51, 66, 139; creolization of, 131–42, 144; in cuisine, 68, 75–80, 92–93, 116–17, 131–34, 151–52, 165, 191–95, 211, 223–31; diversity/varieties, 36–38, 51–52, 141–43; first written account of, 57–58; in Florida, xiv, 26, 196–218, *207*, *209*; flowers of, 26–28; as food crop, 144, 219; genetics of, 33, 51, 137–42, 167–68, 199, 203–5, 220, 222, 225, 231; globalization of, 164–80; historical overview of, 92–105, 111; in literature, 61–67; maritime uses, 20–22, *21*; naming of, xv–xvi, 123–25, 224–25, 229–30; online ordering/shopping, 188–89; picking/harvesting, 24, 26, 28–32, 87, 105, 125, 145–47, *169*, 183, 204–9, *207*; preservation techniques, 13–14, 32; propagation of, 32–35, *34*, 53, 222; reparations and, 119, 226–28; resilience of, 68, 155, 172, 209, 233; sap of, xii, 25, 27, 29–30; seeds of, 24, 28–29, *29*, 33, 53, 94–95, 97, 200–202, 206–8, 213, 222; shelf life of, 12–13, 32; skin of, 27, 32, 81; slavery and, ix, xii–xiii, xiii, 88, 92–94, 116–30, 131, 156, 226–28; superfood claims, xiii, 181–95, 223; sustainability and, xiii, xiv; taste of, xi, 30; texture of, xi, 28, 31–32; vodka from, 191–94; wood, uses for, xii, 19–22, 24–25, 77, *78*, 134, 172–73, 177, 185
breadfruit in Africa: African breadfruit (*Treculia africana*), 210–11; Ghana, 29, 164–70, *169*, 179–80, 194–95, 223; Tanzania, 5, 198; Uganda, 171
breadfruit in Hawai'i: bird liming with, 30; Breadfruit Institute, 34, 167, 219; canoe crops, 45–48, 60, 71, 74, 79, 87–88, 104, 118, 223; Cook, James and, 45–46, 56–57, 76, 98; Hawai'i 'Ulu Cooperative, 80–81, 85; *hoʻokupu* (offering) and, 20–21; *imu* (earthen oven), 231; Kahanu Garden and Preserve, 52–53; Kū (Hawaiian god), 4, 49, 177, 230; Lili'uokalani, Queen, 85–86; *'ohana*, 4; Polynesian Cultural

Center, 118; Polynesian culture, 45, 48, 61, 64–74, 87, 118, 165, 202, 222; Polynesian Voyaging Society, 74; surfboards from, 20–22, *21*, 53, 134, 173, 185; uses and cultivation, *3*, *25*, 28, 32–34, 52, 54, 76–88, 172–73, 191–93, 198, 221–24, 229–31

Breadfruit Institute, 34, 167, 219

breadfruit in Tahiti: barkcloth and, 15, 19; in beer and alcohol, 191, 193; early explorers and, 18, 59–62, 67–77, *78*, 98–105, 123–24, 140, 200; introduction to, xiii–xiv, xv, 5–6; uses and cultivation, *33*, 36–37, *37*, 48–50, 92, 110–12, 125–26, 135, 142, 166, 219, 222, 225–26

breadfruit in the Caribbean, Central America, and Atlantic: Bahamas, *90*, 156–57; Barbados, *25*, *91*, 107, 125, 134, 143–51, *149*, 167; Bermuda, xiv, *25*, 215, 217, *218*, 223; Colombia, 221–22; Costa Rica, 32, 167; Haiti (formerly Saint-Domingue), *90*, 93, 120–21, 132, 171, 200–201, 206; Honduras, 167, 223; Jamaica, *90*, 94, 96–97, 111, 116, 132–35, 141–43, 151–52, 155–60, 165–67, 200–201, 219–22, 225–26, 231; Puerto Rico, *91*, 93, 151–56, *152*, 197, 222; St. Barthélemy, *91*, 128, 151; St. Croix, *91*, 108, 192, 223–24; St. Helena, 96; St. Thomas, *91*, 191–93; St. Vincent and the Grenadines, xiii, *91*, 93–97, 102, 108, 111–12, 116, 122, 128, 131–42, 152, 157–60, 219–21, 225–26

breadfruit in the Pacific: Australia, xiv, 2, 167, 211–14; Borneo, xiv, 40–42, *42*; Chuuk Islands, 45; Guam, xiv, xv, 7, 10, 24, 56, 58, 86–87; Guatemala, 139–40; Mariana Islands, xiv, xv, 24, 74; Marquesas, xiv, xv, 13, *16*, 47–48, 57, 65–66; Micronesia, Federated States of, xiv, xv, 6–13, *11*, 15, 74, 221; New Guinea, xv, 2, 24, 39–44, 52, 166, 198, 219; New Zealand, 2, 45–47, 61–63, 76, 94, 202; Philippines, 24, 56, 59; Solomon Islands, 2, 39, 45, 57; Teti'aroa island, xiv, 67–71. *See also* breadfruit in Hawai'i; breadfruit in Tahiti

Breadfruit Line, 167, 168

breadfruit revolution, 6–7, 10, 12

British acquisition of breadfruit, xii, 75, 92–93, 96–112, 119–20, 124, 129

Bruce, Julia, 112

Bucke, Charles, 112

Buffett, John, 104

Burchard, Sarah, 80

Cadman, John, 53–54

canoe crops, 45–48, 60, 71, 74, 79, 87–88, 104, 118, 223

carbon dioxide (CO_2), 174–75, 177–78

Caribbean. *See* breadfruit in the Caribbean, Central America, and Atlantic

Caribbean Community (CARICOM), 227–28

Caribbean Youth Environmental Network, 167

Carmichael, Abayomi, 217

Cashmore, Ellis, 133

Casid, Jill, 117

celebrity chefs, 79–80

Chang, Fong-Ming, 43

Chee, James, 40

Chi Ching Ching, 159, 219

Choy, Sam, 79–80, 192

Christian, Fletcher, 69, 98–100, 102–3

Christie, Fairlie, 200–201

Chuuk Islands, 45

Clarke, Austin, 145
climate change: breadfruit and, xiii, xiv, 172, 198; carbon dioxide (CO_2), 174–75, 177–78; climate zones and, 196–218; fossil fuels and, 173–75; greenhouse gases, 175, 177–78; impact of, 193
Clinton, Bill, 122
coffee, 76–77, 113–14, 139–40
Cole, Hannah Rachel, 92, 129, 226–27
Coleridge, Samuel Taylor, 67
Colombia, 221–22
colonialism, 62, 95, 119, 128–29, 144, 155, 164, 229
colors of breadfruit, xi, 14, 18–19, 24, 26–27, 51, 66, 139
Cook, James, 5, 17–18, 45, 60–61, 71–72, 76, 98
Costanza, Robert, 175–77
Costa Rica, 32, 167
COVID-19 pandemic, xiii, 83, 143, 158
Coxe, William, 112
Crane, Jonathan, 198
creolization of breadfruit, 131–42
Crop Trust, 53
Crosby, Alfred, 94
Crumbley, Phil, 213–14
culinary fusion, 79–80

Dampier, William, 58–59, 124, 131, 191
Darsy, Eugene, 7–12
Darwin, Charles, 54
Davis, Wade, 46
Dawes, Gavin, 56–57
Dening, Greg, 110
Diamond, Jared, 123
Dirac, Paul, 73
dish *de résistance*, 92–105
Dodd, Robert, 100, *101*
Dodson, Travis, 81–82
Dondero, David, 197

dopamine receptor (DRD4) gene, 41, 43–44
Drori, Jonathan, 25, 230
Drunken Botanist, The (Stewart), 191
Dukes, Jeffrey, 174

East India Company, 96–97
Ednie, Elias, 81
Erman, Michel, 93
Eudell, Demetrius, 120
Evans, John, 104

Fairchild, David, 23
famine, 4, 60, 87, 230
female flowers, 26–27
Fermor, Patrick Leigh, 122, 127
Ferraro, Marie, 215, *216*
Fiegl, Amanda, 187
Finney, Ben, 74
Fisher, Scott, 229
Fitzgerald, Matt, 181
FitzRoy, Robert, 54
Fletcher, Ifan Kyrle, 115
Florida, xiv, 26, 196–218, *207*, *209*
flowers of breadfruit, 26–28
food security, xiii, 83, 146, 153, 167–68, 179, 229
food trends, xiii, 190
Forde, Dwight, 148–51, *149*
fossil fuels, 173–75
Franklin, Benjamin, 123
Fraser, Carmeta, 146
free radicals, 182–83
fruit à pain, xv, 127–28, 151
Fruits of Warm Climates (Morton), 197
Fussell, Paul, 106

García, Francisco, 74–75
Garden of Eden, Biblical, 60, 123, 208, 219–20
Garvey, Marcus, 119–20

Garvey, Patrick, 208–9, *209*
Gasik, Lindsay, 40
Gates, Bill, 178
Gauguin, Paul, 63–64
genetics: of breadfruit, 33, 51, 137–42, 167–68, 199, 203–5, 220, 222, 225, 231; DRD4 gene, 41, 43–44; of people, 41, 43–44, 122, 183
Ghana, *29*, 164–70, *169*, 179–80, 194–95, 223
Gibbons, Andrea, xiii
Giroud, Alexandre, 201–2, 206, 208
Giving Tree, The (Silverstein), 5
Glissant, Édouard, 117
"Glory of Toil, The" (Proctor), 61–62
Gloster, Michael, 157–59
gluten/gluten-free foods, xiii, 32, 82, 146, 160, 184–85, 223
Goebel, Roger, 212
Goodman, Jordan, 111–12
Gorrie, John, 196
Goucher, Candice, 136
Green, Elizabeth, 46
greenhouse gases, 175, 177–78
Grimal, Adolf, 208–9
ground provisions. *See* provisions
Guam, xiv, xv, 7, 10, 24, 56, 58, 86–87
Guatemala, 139–40

Haden, Roger, 75
Haiti (formerly Saint-Domingue), *90*, 93, 120–21, 132, 171, 200–201, 206
Hamblin, Kim, 148–51
Hamilton, Alexander, 17, 108–9
Handy, Craighill, 46
Handy, Elizabeth Green, 46
harvesting breadfruit. *See* picking/harvesting breadfruit
Hawai'i. *See* breadfruit in Hawai'i
Hawai'i 'Ulu Cooperative, 80–81, 85
Hawkesworth, John, 109–10

Heitaa, Félicienne, 13–14
HMS *Bounty*, 67, 98–103, 129, 193, 255n11; *Mutiny on the Bounty* (film), 69–70; *The Bounty* (poem), 129–30
HMS *Providence*, 102, 135, 140–142, 225–26
Hobbes, Thomas, 126
Hoch, James, 57
Hōkūle'a, 74
Hollis, Paul, 217
Honduras, 167, 223
Hooker, Joseph Dalton, 54
ho'okupu (offering), 20–21
Hosack, David, 17
Hough, Richard, 98, 100
Hugo, Victor, 106
Humboldt, Alexander von, 62
Hursthouse, Charles, 61, 62–63

imu (earthen oven), 231
Ines, Harrison, 80
International Treaty on Plant Genetic Resources for Food and Agriculture, 167–68
ital livit, 132–33

jackfruit, 23–24, 191, 205, 211–12
Jacobsen, Sven-Erik, 186
Jamaica, *90*, 94, 96–97, 111, 116, 132–35, 141–43, 151–52, 155–60, 165–67, 200–201, 219–22, 225–26, 231
Jarrett, Denton, 217–18, *218*
Jefferson, Thomas, 200–203, 206
Jelliffe, Derrick, 185
Johnson, Samuel, 124–25
Jungle Project, 167

Kahanamoku, Duke, 20
Kahanu Garden and Preserve, 52–53
Kamel, Georg Josef, 59
Kew Gardens, 54, 96, 97, 111
Kimmerer, Robin Wall, 131, 134

Kincaid, Jamaica, 93, 97–98, 118–19, 121, 225
Kirch, Patrick, 45–47
Kitson, Peter, 66–67
konparet, 171
Kū (Hawaiian god), 4, 49, 230
Kurlansky, Mark, 107–8

Labat, Père, 95
Labouisse, Jean-Pierre, 51
Lammers, Duane, 82, 177
Lapita people, 44–45, 48, 61, 74, 87, 222
Laudan, Rachel, 186–87
Leakey, Colin, 199–200, 203
Let Us Beat Swords Into Ploughshares (Vuchetich), 225–26, *226*
Levy, Pevatunoa, 36–38, *37*
Lewin, Louis, 8
Lewis, David, 73–74
Li, Fabiana, 186
L'île rêves écrasés (The Island of Shattered Dreams) (Spitz), 49–50
Lili'uokalani, Queen, 85–86
Lincoln, Noa Kekuewa, 76–77, 172–73
Livingston, Chad, 172–73
L'Ouverture, Toussaint, 93
Lowenthal, David, 93–94, 117
Lunalilo Home, 83
Lythberg, Billie, 16–17

Magellan, Ferdinand (Fernão de Magalhaes), 56
Maillis, Pericles, 156–57
Maire, Tihone, 68
Makowiecki, Mark, 220–21
Māla Kalu'ulu, 77
male flowers, 26–27
Maloof, Karim Ganem, 222–23
ma (mashed and fermented breadfruit), 13–15, 230
Manley, Todd, 191–94

Māori people, 63
Mariana Islands, xiv, xv, 24, 74
maritime uses of breadfruit, 20–22, *21*, 45–48, 60, 71, 74, 79, 87–88, 104, 118, 223
Marquesas, xiv, xv, 13, *16*, 47–48, 57, 65–66
Mars, Roman, 159
Martes, Jesús, 151–53
Martí, José, 93
Mason, Sandra, 143–45
Matías, Edgardo, 154–55
McDonough, Chris, 174
McLaughlin, Mary, 167, 171
McLaughlin, Mike, 167, 171
McWilliams, James, 187
megafauna, 40
Melville, Herman, 64–66
Mendaña y Neira, Álvaro, 57–58
microclimates, 199, 203, 210
Micronesia, Federated States of, xiv, xv, 6–13, *11*, 15, 74, 221
Milligan, Mark, 223–24
Mitchell, James, 122
Moby-Dick (Melville), 65
Molimau-Samasoni, Seeseei, 79
Moreau, Charles, 151
Morin, Mathieu, 119
Morris, Amanda, 140
Morris, Valentine, 112, 114–15
Morrison, James, 67
Morrissey, Michael, 111
Morton, Julia, 197
Mottley, Mia, 143–44
Mulcahy, Matthew, 107
Musgrave, Toby, 96, 109–12
Myth of the Lazy Native, The (Alatas), 62
myths and stories, 4, 49–50

Naipaul, V. S., 93, 131
naming of breadfruit, xv–xvi, 123–25, 224–25, 229–30

Nan Madol archaeological site, 7–9, *9*
national dishes, 93–94, 125, 228, 231
National Tropical Botanical Garden, 34, 52, 197–98
Newell, Jennifer, 100–101, 109
New Guinea, xv, *2*, 24, 39–44, 52, 166, 198, 219
Newton, Thomas, Jr., 201
New Zealand (Aotearoa), *2*, 45–47, 61–63, 76, 94, 202
Nixon, Richard, 122
Nowell, Charles, 55

Obama, Barack, 67
Oduro, John Jurai, 194–95
'ohana, 4
One Village Farm, 81–82
Opgenorth,, Mike, 52–53
Oster, Gerald, 120–21
Oster, Selmaree, 120–21

Pacific Agribusiness Research for Development Initiative, 167
Paloma, Diane, 83
pana, 152–55, 188, 193
Papuan people, 39–44, 124. *See also* New Guinea
Parry, John, 116
Pauly, Daniel, 136
Peter Pan (Barrie), 64
Petersen, Gary, 6
Philippines, 24, 56, 59
Piailug, Mau, 74
picking/harvesting breadfruit, 24, 26, 28–32, 87, 105, 125, 145–47, *169*, 183, 204–9, *207*
plant-based foods, xiii, 133
Pōhaku (Tom Stone), 20–21, *21*
Pohnpei Historic Preservation Office, 7
Pollan, Michael, 85, 182
Polynesian Cultural Center, 118

Polynesian culture, 45, 48, 61, 64–74, 87, 118, 165, 202, 222
Polynesian Voyaging Society, 74
Pono Pies, 53–54
popoi, 13–15, *16*, 64–65
pōpolo, 223–24
Prendergast, Ellen Keko'aohiwaikalani Wright, 86
preservation techniques, 13–14, 32
Principles of Quantum Mechanics, The (Dirac), 73
Proctor, Edna Dean, 61–62
propagation techniques, 32–35, *34*, 53
Protected Designation of Origin laws, 138
Providence. See HMS *Providence*
provisions, ix–xiii, 18, 47, 70, 74, 100, 107, 111, 114, 120, 160, 222
Puerto Rico, *91*, 93, 151–56, *152*, 197, 222
purau leaf, 13–14

Queirós, Pedro Fernandes de, 57–58
Quinn, Daniel, 123

Ragone, Diane, 5, 34, 52, 186
Ramsay, Gordon, 79–80
Rare Fruit Council, 203–4
Rastafarian movement, 132–33
Reindorf, Carl Christian, 165
reparations, 119, 226–28
resilience of breadfruit, 68, 155, 172, 209, 233
Roatta, Rane, 189
Roberts-Nkrumah, Laura, 137
Rohlehr, Gordon, 117
Rollins, Chris, 209–10, 213
Rousseau, Jean-Jacques, 123

Sacks, Oliver, 7
Sahlins, Marshall, 126
Saint-Domingue. *See* Haiti

sakau, 7–9, 12, 21, 230
Salleras, Peter, 211–12
Samuel, John, 99
Santa Cruz Islands, 45
Santangelo, Rafael, 204–5
sap of breadfruit, xii, 25, 27, 29–30
Schiebinger, Londa, 95
Schlegel, Edelle, 189–90
Scientific Research Organisation of Sāmoa (SROS), 79
SDGs. *See* UN Sustainable Development Goals
seeds of breadfruit, 24, 28–29, *29*, 33, 53, 94–95, 97, 200–202, 206–8, 213, 222
Shapiro, Dana, 77
Shark Going Inland Is My Chief, A (Kirch), 46–47
Sharp, Andrew, 72
Shaw, Alexander, 17–18
shelf life of breadfruit, 12–13, 32
Shoal of Time (Dawes), 56–57
Silverstein, Shel, 5, 222
skin of breadfruit, 27, 32, 81
"slave food," 92–94, 128–29, 131, 156
Slavery Abolition Act (1833), 119, 120
slavery and breadfruit, ix, xii–xiii, 88, 92–94, 116–30, 131, 156, 226–28
slavery reparations, 119, 226–28
Sloan, Hans, ix
Smilowitz, Erika, 128
Smith, Adam, 177
Smith, Philippa Mein, 202
Smith, Tammy Mahealani, 82–85
Solomon Islands, 2, 39, 45, 57
Spary, Emma, 110
sphagnum moss, 34
Spitz, Chantal, 49–50
St. Barthélemy, *91*, 128, 151
St. Croix, *91*, 108, 192, 223–24
St. Helena, 96
St. Thomas, *91*, 191–93

St. Vincent and the Grenadines, xiii, *91*, 93–97, 102, 108, 111–12, 116, 122, 128, 131–42, 152, 157–60, 219–21, 225–26
St. Vincent Botanical Gardens, 97, 157
Steel, Leith, 190–91
Stevenson, Robert Louis, 63–64
Stewart, Amy, 191
Stone, Tom (Pōhaku), 20–21, *21*
Stuart, Andrea, 122
superfood claims, xiii, 181–95, 223
Support Roatán, 167
surfboards, 20–22, *21*, 53, 134, 173, 185
sustainability and breadfruit, xiii, xiv, 54, 82, 166–85, 190, 193–94, 221–23. *See also* UN Sustainable Development Goals
Sustainable Development Goals (SDGs). *See* UN Sustainable Development Goals
Svalbard Global Seed Vault, 53
Swan, Charles, 58

Tanzania, 5, 198
taste of breadfruit, xi, 30
Teavai-Murphy, Hinano, 26–27, 168
Teti'aroa island, xiv, 67–71
texture of breadfruit, xi, 28, 31–32
Thicknesse, Philip, 115
Thirty Meter Telescope (TMT) protests, 84–85
Thompson, Jerrol, 157
Tree of Knowledge of Good and Evil, 219–31
Tree of Life, 219–31
Trees That Feed Foundation, xiii, 166–67, 170–71
Turner, Jack, 81
Tutu, Desmond, 134
Twenty Thousand Leagues Under the Sea (Verne), 66

Typee: A Peep at Polynesian Life (Melville), 64–66

Uganda, 171
uhm (earthen oven), 9–12, *11*
Ulloa, Antonio de, 135
'ulu. See breadfruit
UN Sustainable Development Goals, 167, 170–72, 175

Vaughan, Benjamin, 200
vegetarianism, xiii, 80, 133, 148
Verne, Jules, 66
Villalobos, Marisol, 151, 153, 219
vodka from breadfruit, 191–94
Voltaire, 113–15
Vuchetich, Evgeniy, 225

Walcott, Derek, 121, 128–30
Wallace, Alfred Russel, 187
Walsh-Dilley, Marygold, 186

Washington, George, 200
Watts, David, 117
Wealth of Nations, The (Smith), 177
Wesley, Charles, 94
Westlake, Wayne Kaumualii, 118
White, Paul, 110
Whitman, Bill, 203–4, 205, 208
Wiles, James, 135
Williams, Evelyn, 40
Williksen-Bakker, Solrun, 50–51
Wilson, Woodrow, 57
Wims, Sam, 206, *207*
Wolfe, Herbert S., 199
wood, uses for breadfruit, xii, 19–22, 24–25, 77, *78*, 134, 172–73, 177, 185
World Health Organization, 75–76

Zaldivar, Jorge, 204–5, 213, 215
Zerega, Nyree, 140
zone-pushing, 199, 203, 206, 209–10, 217, 223

GPSR Authorized Representative: Easy Access System Europe, Mustamäe tee
50, 10621 Tallinn, Estonia, gpsr.requests@easproject.com

www.ingramcontent.com/pod-product-compliance
Lightning Source LLC
Chambersburg PA
CBHW022038290426
44109CB00014B/903